Cybersecurity Ethics

This textbook offers an accessible introduction to the topic of cybersecurity ethics. The second edition has been revised and updated, and contains new chapters on social justice, AI, and Big Data.

The book is split into three parts. Part I provides an introduction to the field of ethics, philosophy, and philosophy of science, three ethical frameworks – virtue ethics, utilitarian ethics, and communitarian ethics – and the notion of ethical hacking. Part II applies these frameworks to particular issues within the field of cybersecurity, including privacy rights, surveillance, and intellectual property. The third part concludes by exploring current codes of ethics used in cybersecurity, with chapters on artificial intelligence, social diversity, Big Data, and cyberwarfare. The overall aims of the book are to:

- Provide ethical frameworks to aid decision-making
- Present the key ethical issues in relation to computer security
- Highlight the connection between values and beliefs and the professional code of ethics

The textbook also includes three different features to aid students: "Going Deeper" features provide background on individuals, events, and institutions in cybersecurity; "Critical Issues" features contemporary case studies; and "Tech Talks" contain features that assume some familiarity with technological developments.

The book will be of much interest to students of cybersecurity, cyberethics, hacking, surveillance studies, ethics, and information science.

Mary Manjikian is Professor at the College of Arts and Sciences, Regent University in Virginia Beach, USA, and a professor in the Fleet Seminar Program for the US Naval War College. She is the author of several books, including, most recently, *Introduction to Cyber Politics and Policy* (2020), and is co-editor of the *Routledge Companion to Global Cyber-Strategy* (2021).

'This book is a rich and valuable resource for students to learn more about non-technical aspects of cybersecurity.'

—Markus Christen, *Managing Director, Digital Society Initiative, University of Zurich, Switzerland*

'The authoritative text on cybersecurity ethics, covering a wide selection of topics including privacy, intellectual property theft, artificial intelligence and big data. This is a terrific resource for students in Europe and beyond.'

—Max Smeets, *ETH Zurich, Center for Security Studies (CSS) and Director of European Cyber Conflict Research Initiative*

'A comprehensive tour de force in cybersecurity ethics. This illuminating book weaves together ethical frameworks that address the evolving nature of some of the most complicated cyber issues in the 21st century. Are we prepared to understand and respond to the ethical challenges that continue to emerge across cyber domains and technological advances? The author addresses these challenges head on and this book will prove to be an excellent resource to those who wish to advance their understanding of cybersecurity ethics.'

—William Travis Morris, *Director, Norwich University Peace and War Center, USA*
—Norwich University

Praise for the 1st edition

'This book is a bold and innovative synthesis of thinking from diverse yet interlinked disciplines. It is vital reading for scholars, policymakers, security professionals and organizational leaders. Manjikian's explication of the ACM Code of Ethics shows why it is a foundational concept for cybersecurity.'

—Steven Metz, *U.S. Army War College, USA*

'As cyber conflict, espionage and crime increasingly challenge nations and their citizens, Manjikian's *Cybersecurity Ethics* provides a comprehensive and needed addition to the cyber literature canon. This work constitutes a robust framework for decisions and actions in cyberspace and is essential reading for policymakers, practitioners, and students engaging the field of cybersecurity.'

—Aaron F. Brantly, *Army Cyber Institute, United States Military Academy, West Point*

'Mary Manjikian's introduction to cybersecurity ethics nicely links philosophy to practical cyber concerns of students, corporate and government information managers, and even cyber warriors. Complicated concepts are easy to understand and relevant to personal decision-making.'

—John A. Gentry, *Georgetown University, USA*

'Dr Manjikian has done a masterful job of outlining ethical standards to the constantly evolving cybersecurity domain. This book is a vital reference for those who are concerned with ethics related to hacking, privacy, surveillance, and cyberwarfare in an ever changing virtual environment that transcends boundaries and cultures and challenges the traditional ways that humans have dealt with each other. Ground-breaking and should be required reading for any serious cybersecurity professional.'

—Keith Dayton, *George C. Marshall European Center for Security Studies, Germany*

'A great introductory text to complex conceptual and practical issues in cybersecurity.'

—Heather Roff, *Arizona State University, USA*

'Mary Manjikian's excellent textbook, *Cybersecurity Ethics*, delivers an eminently readable perspective for the experienced professional as well as a casual cyberspace consumer. . . . Overall, *Cybersecurity Ethics* excellently fulfills the stated goal to provide an ethical framework and illustrate several current issues regarding cybersecurity through a new textbook. The construction was splendid, material assembled coherently, and sources presented thoroughly and accurately. This book is an excellent introduction to ethical processes associated with cyberspace.'

—Mark Peters, *Strategic Studies Quarterly*

Cybersecurity Ethics

An Introduction

Second Edition

Mary Manjikian

LONDON AND NEW YORK

Cover image: © Getty Images – Credit: monsitj

Second edition published 2023
by Routledge
4 Park Square, Milton Park, Abingdon, Oxon OX14 4RN

and by Routledge
605 Third Avenue, New York, NY 10158

Routledge is an imprint of the Taylor & Francis Group, an informa business

First edition published by Routledge 2017

British Library Cataloguing-in-Publication Data
A catalogue record for this book is available from the British Library

Library of Congress Cataloging-in-Publication Data
Names: Manjikian, Mary, author.
Title: Cybersecurity ethics : an introduction / Mary Manjikian.
Description: Second edition. | Abingdon, Oxon ; New York, NY : Routeldge, 2023. | Includes bibliographical references and index.
Identifiers: LCCN 2022036730 (print) | LCCN 2022036731 (ebook) | ISBN 9781032164991 (hbk) | ISBN 9781032164977 (pbk) | ISBN 9781003248828 (ebk)
Subjects: LCSH: Internet—Moral and ethical aspects. | Computer networks—Security measures—Moral and ethical aspects. | Computer crimes.
Classification: LCC TK5105.878 .M36 2023 (print) | LCC TK5105.878 (ebook) | DDC 005.8—dc23/eng/20221017
LC record available at https://lccn.loc.gov/2022036730
LC ebook record available at https://lccn.loc.gov/2022036731

ISBN: 978-1-032-16499-1 (hbk)
ISBN: 978-1-032-16497-7 (pbk)
ISBN: 978-1-003-24882-8 (ebk)

DOI: 10.4324/9781003248828

Typeset in Bembo
by Apex CoVantage, LLC

To my husband Ara, whose paranoia about everything internet teaches us to be more careful.

Contents

Tables

Figures

Boxes

Acronyms and Abbreviations

ACLU:	American Civil Liberties Union
API:	Application Programming Interface
AWS:	autonomous weapons systems
BCI:	Brain-Computer Interface
BWC:	Body-worn camera
CVE:	Common Vulnerabilities and Exposures database
DNN:	Deep neural network
CEH:	Certified Ethical Hacker
CFAA:	United States Computer Fraud and Abuse Act
CISA:	United States Cybersecurity and Infrastructure Security Agency
CREST:	Council of Registered Ethical Security Testers
DHS:	United States Department of Homeland Security
DLT:	Distributed Ledger Technology
DOD:	United States Department of Defense
DPIA:	Data Protection Impact Assessment
DTC GT:	Direct to Consumer Genetic Testing
EEOC:	United States Equal Employment Opportunities Commission
EPM:	Electronic Performance Monitoring
EC-Council:	International Council of E-commerce consultants
EU GDPR:	European Union General Data Protection Regulation
ECPA:	the United States Electronic Communications Privacy Act
EI-ISAC:	United States Elections Infrastructure Information Sharing and Analysis Center
FERPA:	United States Family Education Rights and Privacy Act
FISA:	the United States Foreign Intelligence Surveillance Act
GCHQ:	United Kingdom Government Communications Headquarters (similar to the US National Security Agency in its functions)
GLBA:	United States Gramm Leach Bliley Act (GLBA)
HBCU:	Historically Black Colleges and Universities
HIPAA:	United States Health Insurance Portability and Accountability Act
HHS:	United States Department of Health and Human Services
IACC:	International Anti-Counterfeiting Coalition
ICANN:	Internet Corporation for Assigned Names and Numbers
ICE:	United States Immigration and Customs Enforcement
IEEE:	Institute of Electrical and Electronics Engineers
IFPI:	International Federation of the Phonographic Industry

IRB:	Institutional Review Board
ISACA:	Information Systems Audit and Control Association
ISSA:	Information Systems Security Association
LBS:	location-based services
LMS:	learning management system
LPPM:	Location-Privacy Preserving Mechanisms
MHC:	meaningful human control
MLAT:	Mutual Legal Assistance Treaty
NDA:	Non-Disclosure Agreement
NFT:	Non-Fungible Token
NIST:	United States National Institute on Standards and Technology
NSA:	United States National Security Agency
NVD:	United States National Vulnerability Database
PET:	privacy-enhancing technologies
PIA:	Privacy Impact Assessment
RFID:	Radio Frequency Identification
SaaS:	software as a service; surveillance as a service
SCR:	Semantic Cloaking Regions
SDN:	Specially Designated Nationals and Blocked Entities
SEC:	US Securities and Exchange Commission
SOPA:	Stop Online Piracy Act
TIDE:	US Terrorist Identities Datamart Environment
TOM:	Theory of Mind
TOS:	Terms of Service
TRIPS Agreement:	The Agreement on Trade-Related Aspects of Intellectual Property Rights
TSDB:	Terrorism Suspects Database
UI:	User Interface Design
UX:	User Experience Design
UNHCR:	United Nations High Commissioner for Refugees
WHO:	World Health Organization

Preface to the Second Edition

Welcome to the second edition of *Cybersecurity Ethics: An Introduction*. This textbook first appeared in 2017, and I was inspired to write it due to my dissatisfaction with existing undergraduate and graduate texts in cybersecurity ethics. At that time, I was aware that those who teach cybersecurity ethics come from various backgrounds – some from academic computer science or engineering, some from policy studies, and some from the military. Some are philosophers by training, while others are practitioners in the computer industry. Those who are not philosophers struggle to master what may seem like arcane and abstract philosophical debates about the nature of existence, being, and agency. In contrast, those who are academic philosophers may struggle to understand and apply technical concepts like Location-Based Privacy Mechanisms and Non-Fungible Tokens (NFTs).

Thus, I have created a resource that appeals to this broad spectrum of professors and students. This textbook assumes neither prior study in philosophy nor a high level of technical knowledge. I hope that in providing this resource, I have created a place where those who share an interest in the philosophy of technology, ethics, and cybersecurity can begin to have conversations across disciplinary boundaries – so that philosophers may talk with policy experts, lawyers, and engineers as we work together to model, create, and practice ethical behavior in cyberspace.

An ethics course does not aim to teach you precisely what you should do in every ethical dilemma you will encounter professionally. However, as much as possible, this text focuses on real-world situations you might encounter in your day-to-day work. Part I of this text (Chapters 1, 2, and 3) is the most abstract, but you will rely on the concepts introduced here throughout the course.

This text contains ten chapters: Chapter 1 introduces the field of ethics, describing how it differs from either law or religion and why it is still necessary when we have both law and religion. It lays out some of the differences and debates between philosophers as well. Chapter 2 introduces three stances that can be applied to thinking about ethics: virtue ethics, utilitarianism, and deontological ethics. Chapter 3 introduces the notion of ethical hacking, the hacker code, and the particular problem of penetration testing. Chapter 4 examines ethical issues related to privacy, while Chapter 5 looks at surveillance practices. Chapter 6 considers the problem of piracy and intellectual property theft. Chapter 7 introduces ethical issues related to our ever-increasing reliance on artificial intelligence. In contrast, Chapter 8 considers ethical issues related to diversity, equity, and inclusion (DEI) due to that reliance on artificial intelligence. Chapter 9 considers Big Data and ethical issues related to cybersecurity and Big Data. We conclude in Chapter 10 with a look at some emerging ethical issues in the field of military cybersecurity, again looking at developments in the field of artificial intelligence.

The book also contains three different types of added features. "Going Deeper" features provide background on individuals, events, and institutions in cybersecurity. "Critical Issues" features look at stories you might see in the news – from the ethical issues encountered in working with cryptocurrency and NFTs to issues related to new types of surveillance like Body-Worn Cameras (BWCs) and Smart Cities. Finally, "Tech Talks" contain features that assume some familiarity with technological developments. These features allow those interested in, for example, privacy, to think about specific new types of tools like semantic cloaking, which they might wish to use in their work. My website, marymanjikian.com, also contains some additional materials for those who are more interested in applied technical features related to cyber ethics, including additional discussion questions and resources by chapter.

For those who may have used the first edition of this text, you will notice some changes in this new edition. I am grateful to the anonymous reviewers who responded to a request for feedback and for the lists of features which they requested that I include in a new edition. In response to their suggestions, I have included new material on neuro-privacy and brain hacking as part of the Chapter 4 discussion on privacy. Chapter 5's surveillance discussion now includes new material on "surveillance capitalism," including how workers encounter surveillance and how surveillance can occur in the "smart cities" of the future.

As technology progresses and becomes "democratized," or more readily available to all, including those with a nontechnical background, we will increasingly see nonstate actors posing threats to national and commercial cybersecurity. Thus, the material on disinformation, ransomware, and cryptocurrency pose questions about the roles played by nonstate actors.

A brand-new Chapter 7 introduces definitions of artificial intelligence and autonomy and raises new ethical issues associated with the agency humans retain and the agency and autonomy granted to machines. Here you will encounter the ethical notion of meaningful human control (MHC).

As the world faced the COVID pandemic, cybersecurity practitioners and scholars have begun asking new questions about health data and how health data privacy can be preserved. These questions are dealt with in special insets on COVID and intellectual property, health data privacy, and health data surveillance practices.

As workplaces and societies have begun paying renewed attention to practices of diversity, equity, and inclusion in the wake of movements like the Black Lives Matter movement, cybersecurity practitioners have begun to consider the ethics of Big Data and machine learning. How can we ensure that fairness, access, and equity are upheld? These issues are examined in Chapter 8, which asks questions about diversity, equity, and inclusion in relation to cybersecurity.

The role of the cybersecurity practitioner has also changed with the advent of Big Data and artificial intelligence. Some more routine cybersecurity tasks may now be automated and outsourced to AI. Chapter 7 on artificial intelligence and Chapter 9 on Big Data allow cybersecurity practitioners to consider how data analytics can be used to understand patterns of cybersecurity incident breaches better and to think about how their jobs now intersect with those of data analysts and architects – as decisions about data pipelines and data storage have real-world impacts on our ability to secure our data.

Finally, as you will see throughout this volume, the meaning and task of cybersecurity have expanded exponentially with the advent of the Internet of Things. Protecting individual, corporate, and government information and data from intrusion, theft, and destruction has become more challenging as data and information flows through our connected homes, devices, and even wearable technology! While once a cybersecurity practitioner may have conceptualized her job as protecting the data that resides within the corporation that employs her, today her task may be much more complex – encompassing considerations like the data "exhaust" that employees produce as they go about their daily lives, the overlap between personal and professional data – particularly when people are working from home – and considerations like how best to store and structure data so that it is both secure and accessible. Throughout this volume, you will see discussion questions and activities that allow you to engage with complex scenarios where they may be multiple competing values and interests to consider.

While no one text can be all things to all people, I hope that you will benefit from the quick looks provided at such diverse topics as NFTs and intellectual property and the ethics of cryptohacking and cryptojacking.

As always, I welcome your feedback regarding how this text can be improved and made the most relevant to practitioners and analysts in the exciting field of cybersecurity!

> Instructors and students are also encouraged to visit my personal website, marymanjikian.com, where you will find updated features related to stories that are in the news. There you may also find PowerPoint resources to accompany each chapter. You may also contact me to share feedback about the book, which I will consider for future updates. The website also contains a discussion forum where you can connect with others using this text.

Acknowledgments

One of the joys of writing a textbook used by students worldwide lies in the connections and friendships I have made as a result. I am grateful to the many professors and students who contacted me after first encountering my ideas in this textbook and a second book (*Cyber Politics and Policies: An Introduction*) which I have written since 2017. Even during the long, isolated days of the COVID quarantine, I could Zoom with students at the US Air Force Cyber College, the International Relations Club at the University of Delhi, India, and the graduate students in cybersecurity at Marymount University. I appreciate the teachers and professors who contacted me, including those teaching AP Computer Science at high schools in the United States. As a result of these communications, I could better picture my audience and think about which features would be most beneficial to you as you encountered the exciting field of cybersecurity.

I am also grateful to the Digital Issues Discussion Group (DIDG) members, ably coordinated by Professor Christopher Whyte at Virginia Commonwealth University. The ability to communicate with peers worldwide and to be exposed to the latest research in cybersecurity has been inspiring and encouraging. I am also grateful to Scott Romaniuk and the many international researchers in cybersecurity whom we have met as we have collaborated on cybersecurity and artificial intelligence research projects.

Finally, as always, I wish to thank my editor at Routledge, Andrew Humphrys, for his patience and good humor as I struggled with this and other projects.

All errors in this text are, of course, my own.

Part I

1 What Is Ethics?

LEARNING OBJECTIVES

At the end of this chapter, students will be able to:

1. Define philosophy and describe its aims as a field of inquiry
2. Define ethics and cybersecurity ethics and give at least three examples of practical ethical dilemmas that cybersecurity professionals encounter
3. Describe the role of computer engineers in affecting political, social, and economic life through making engineering decisions that have ethical consequences
4. Define significant terms associated with the study of ethics
5. Describe the relationship between ethics, religion, and laws

What Is Ethics?

Turn on the nightly news, and chances are you will hear the word *ethics* bandied about. You might hear references to a parliamentary or congressional ethics inquiry, or you might hear about a public official being brought up on ethics charges. You might associate ethics with law enforcement, lobbying, corruption in government, or investigations related to a person's character, marital fidelity, or predisposition to accept bribes. You might think that ethics are relevant only to lawyers and public officials. But what does ethics, then, have to do with cybersecurity? Although these examples involve current events and law enforcement, ethics is a broad academic discipline with historical roots, encompassing much more than what you see on the nightly news.

To begin our study, we define ethics and consider its academic origins and ways of studying ethics. Ethics is a branch of **philosophy**, an academic subject concerned with the fundamental nature of knowledge, reality, and existence. Within philosophy, ethics considers people's values and sources. It also considers whether they differ over time and place and how people and groups translate those values into behavior. Some values that philosophers consider include justice, equality, and human rights.

You may have encountered empirical methodology in the sciences or social sciences in many of your other academic courses. In both the hard and soft sciences, analysts attempt to observe and measure things that exist in reality (from molecules to patterns of immigration) to formulate and test rules that describe and predict likely results. They ask *empirical questions*, or questions based on measurable and observable observations. These questions might have a yes or no answer or an answer which is a number. Thus, a social scientist like a sociologist might consider evidence that income inequality increases in the United

DOI:10.4324/9781003248828-2

States. He might ask whether or not there are more significant gaps in the United States between wealthy people and those who are poor than there have been in the past, and the causes for that shift.

In contrast, an ethicist, a moral philosopher, considers *normative questions* – questions that do not ask merely about what occurred, how often it occurred, or about the size of a phenomenon – but questions that ask what we as humans *should do* in response to a phenomenon. Thus, an ethicist concerned with income inequality would ask whether people are less generous or community-minded than they were in the past, why that is, and how one could encourage generosity. Both the ethicist and the economist study income inequality, but their methods, assumptions, and research questions differ. Audi describes ethics as "the philosophical study of morality." He tells us that it asks a question like "what ends we ought, as fully rational human beings, to choose and pursue and what moral principles should govern our choices and pursuits" (Audi, 1995, pp. 284–285).

Where Do Ethics and Values Come From?

Do humans have an innate capacity to consider questions of right and wrong? Are there particular acts that every culture finds morally repugnant (like torture or infanticide) and others that all cultures regard as morally acceptable or worthy of respect? Are there universal values that all ethicists and cultures should view as worthy of pursuing, and can these values provide the foundational assumptions for ethical theorizing?

Philosophers called **objectivists** see the ethics project as identifying the right thing to do or the right action to take morally. They believe it is possible to identify solutions to ethical problems using reasoning and ethics tools. They believe that answers to ethical problems can be identified and enacted.

The Christian theologian Thomas Aquinas, who lived in the thirteenth century, referred to *natural law* as describing how values emerge. He believed that core values are found in all human beings, such as the idea that all human life is worthy of respect. He believed a creator had placed these "laws" into every individual. Similarly, the school of ethics known as *Divine Command theory* proceeds from the assumption that a God who exists provides an objective set of ethical standards and that humans can behave ethically if they treat the fulfillment of these standards as their duty (Austin, n.d.; Brackman, n.d.).

However, one does not need to be religious to be an objectivist. In reaching back to Ancient Greece, we can point to Plato's *Theory of Forms* – which suggests that we have an ideal "something" in mind and that we can then compare a particular experience to that ideal. For example, if asking if something is beautiful, we compare it to an ideal standard of beauty that we hold and can access in our minds. He argued that because we can all thus envision beauty or happiness as an ideal, there must be universal values that we all hold to, which could provide the basis for some universal ethical values and recommendations.

However, not all ethicists are objectivists. A *moral relativist* believes that there is no one absolute right or wrong position. Instead, they argue that your subjective perceptions may determine the most moral thing to do. Not everyone will see a moral problem in the same way. Some of us may emphasize one facet of a problem, while others emphasize something different. A moral relativist might thus refute Plato by arguing that beauty standards can differ from one culture to another. Some relativists believe that our perception of ethical behavior depends on social conventions, which vary from society to society. They point to the fact that less than 500 years ago, many people found slavery socially acceptable. While it was still objectively wrong, many people did not yet see it that way. And they point out that virtues considered very important in Victorian England – like chastity or piety – are less important today in many parts of the world. (We should note here that this relativistic view is more commonly found among social scientists who write about ethics than among philosophers who practice ethics.)

Critics of this approach often use moral relativism negatively, suggesting that it "lets people off the hook" or does not hold them morally accountable for holding morally repugnant positions. They argue that we shouldn't overlook a culture's abuse of women or gay people by stating that "according to their culture's understanding at this time, such behavior is moral," for example. What's wrong is wrong, an objectivist would say.

Some analysts in cybersecurity ethics might invoke a moral relativist lens in describing how cultures think differently about concepts like intellectual property in the information technology field. Chang argues that Asian nations have a collectivist culture that prizes equality and the group's progress as a group. In contrast, western nations are more individualistic, prizing individual equity and fairness more than equality and group goals. She says that in Asian cultures, people might regard their activities as sharing rather than theft. She worries that any international code of information ethics that outlawed information sharing might seem like a foreign code imposed on Asian and developing nations. It would be difficult to get people to adhere to that code (Chang, 2012, p. 421).

Ethicists also disagree about whether ethical values change over time. Moral relativists argue that as technologies advance, ethical values can change too. For example, they argue that many people today enjoy sharing on social media, and therefore, people don't value privacy the same way they did in the past. But objectivists argue that ethical decisions rest on stable core values, regardless of one's environment (Calman, 2004). Here we can consider a simple example: Doctors often had a paternalistic relationship with patients in the past. The doctor was an authority who informed the patient of his diagnosis, his health prospects (prognosis), and his treatment. But today, many patients come to the doctor's office having researched their condition online. They already know their diagnosis, prognosis, and treatment options. Here a relativist might identify the emergence of a more cooperative medical ethics model. But an objectivist would note that "old" and "new" doctors still share a common value – caring for and respecting patients. Thus, they would argue, the ethics have not changed, even in a new environment, because the core values endure.

In this text, we present both objectivist and relativist arguments for you to understand the range of ways in which people have considered cybersecurity ethics. However, the main document we refer to throughout this text is the Association for Computing Machinery's (ACM) Code of Ethics. This document presents the values and ethics of the world's largest professional association for computer scientists. This Code of Ethics is universally applicable and internationally relevant to all computer scientists. It is not applied or interpreted differently in different cultures. This professional organization meets from time to time to update this code, which might be modified in the future. However, it presents objectivist, not relativist, ethics.

The First Ethicists

Although we introduced ethics by discussing income inequality questions, ethics don't just consider wealth and poverty. In Ancient Greece, philosophers like Plato and Aristotle asked what constitutes a good life. How can one live a good life – individually and collectively, in concert with others?

Plato (428–348 BC) first asked questions about the virtues or values people should cultivate in their lives. These arguments appear in his published works, most of which are Socratic dialogues. Socratic dialogues are a method of inquiry in which people come together in discussions, posing questions to arrive at new knowledge. He believed that human well-being (or *Eudaimonia*) should be the goal of life and that individuals should cultivate and practice excellence (*arête*) to achieve it. He believed that individuals who lived well-ordered lives could work together to create a well-ordered society. He rejected the idea that people should retaliate against others who harmed them and argued that people were harmed in their souls when they committed injustices. Plato argued that individuals needed to be disciplined in their lives, sacrificing individual wants and needs to create a good society.

While Plato is a well-known western philosopher, ethical thinking has always been global. Chinese thinking about information ethics draws upon Confucius, a scholar who lived in approximately 500 BC. Chang identifies loyalty, duty, and respect for the ties one cultivates with one's community as values central to Chinese culture (Chang, 2012). African ethical thinking rests on tribal values like Ubuntu or concern for harmony within society and groups (Chasi, 2014). All cultures ask questions about justice and equity, conflict and cohesion, but they may answer these questions differently.

As these examples show, ethics is a label for two different ideas: rules for how one should behave or act in the world, and rules for how one should regard the world, oneself, and one's place in it and others. Ethics are thus both rules for action and statements about people's attitudes, what they should love or value, or what they consider necessary.

Because ethical philosophy is normative (asking questions about what one should do) rather than empirical (collecting observations from what can be observed and measured), ethics thus rests on foundational assumptions about what is valuable and essential, which cannot be empirically proven or disproven. An *assumption* is a starting point for an argument taken as a given. The theorist may not require proof of his assumptions but instead may accept that something is true or behave as accurate. For example, a journalist may adopt a journalistic code that states that the journalist's most significant responsibility is to produce authentic or accurate news. A doctor may adopt a medical ethic, stating that his most significant responsibility is to his patient. His responsibilities include curing his patient, alleviating his pain, and treating him respectfully. In considering ethics and ethical decisions, the values on which an ethic rests may be explicit or clearly stated, or merely implied. In evaluating different courses of action and deciding which one is the most ethical, it is thus essential to be aware of the values and assumptions on which the ethic rests, whether the author you are reading spells these out directly or not.

Box 1.1 Tech Talk: The Ethics of User-Centered Design

User-centered design refers to a design process centered explicitly around the needs and experiences of a product's users. (User-centered design can be implemented in designing both user interfaces and user experiences). Rather than building a product and then thinking about how users might interact with it, product engineers who implement user-centered design consider the user's experience and need at every stage of the design process. Rather than modifying a product later if users encounter difficulties using it, designers aim to create a product that works for users from the beginning stages of the design process.

Designing for Users

Specialists in user experience (UX) often have training in psychology and the social sciences – like anthropology. They work to develop mock-ups or prototypes of planned innovations (such as a new user interface for an app running on a smartphone) and then conduct focus groups with potential users. A UX specialist might observe how the user interacts with the product in these focus groups. Do they find it easy to perform tasks on the product, and are there other tasks they struggle with? In developing a product using user-centered design principles, the developers might go through several design iterations – changing fonts, the placement of display buttons, or rewriting any text prompts that accompany the product.

Sometimes we are fortunate enough to encounter a new product that seems remarkably easy to use. We know intuitively how to interact with the device and do not experience a sharp learning curve in getting up to speed. This does not merely happen randomly. Instead, it results from many design refinements and modifications and the application of user-centered design principles.

Schreuder et al. (2013) note that creating a new product should be aimed at a particular end user. In this way, a product can be adapted to be usable by the product's audience. The user-centered design also ideally includes thinking about those who might struggle to use technology – such as those with visual or auditory impairments or other physical disabilities.

Ethics of User-Centered Design

User-centered design thus rests on the ethical principle of empathy since the designer attempts to see the situation from the user's perspective. The designer might think about those who will use a product – including technology-phobic users, older users, those with a disability involving motor skills, or those who are visually or hearing impaired. They also think about the situations in which a tool might be used – for example, someone who uses a phone to call for help after an accident might be frightened and unable to focus quickly. What sort of design would work best in that situation?

User-centered design also rests on the ethical principle of respect. Designers solicit feedback from prospective users of an app or product at all stages of the design process. In doing so, they make several assumptions that respect users. First, they assume that a user who encounters a problem using technology is not slow or stupid. Instead, they accept that the developer is responsible to the user not to waste their time by making something difficult to use. Further, they assume that a user can offer valuable feedback to a developer. (They are not merely an impediment to be dealt with.) While a user may not know how to achieve a particular result, a user may offer valuable feedback about the situations in which a product might be used or the steps required to solve a problem.

Combatting Disparate Impact

One of the goals of user-centered design is thus to identify those who might be marginalized or excluded from using technologies as they exist in their present form. For example, Korngiebel describes how those most likely to visit a health care facility or interact online with health care professionals are often the elderly. People who lack up-to-date technical skills and may also have impairments that make interacting with technology difficult (such as issues with their eyesight or hearing) are often the most dependent on that technology! Therefore, she suggests that in building platforms where users might, for example, access their medical laboratory test results, developers should test these platforms on users such as the elderly (Korngiebel, 2021).

As this example shows, ethical principles are not merely present in abstract philosophical discussions. Instead, they may be applied in everyday situations that developers and engineers encounter daily.

Bibliography

Korngiebel, D. (2021, February 25). Digital healthcare disparities. *Hastings Center Report*. Retrieved June 1, 2022, from https://doi.org/10.1002/hast.1208.

Lowdermilk, T. (2013). *User-centered design: A developer's guide to building user-friendly applications*. Sebastopol, CA: O'Reilly Media.

Schreuder, M., Ricciol, A., Risetti, M., Dahne, S., Ramsay, A., Williamson, J., & Mattia, D. T. (2013). User-centered design in main-computer interfaces – A case study. *Artificial Intelligence in Medicine, 59*(2), 71–80.

Choosing Between Ethical Values

Moving from a purely abstract or philosophical discussion of ethics to a more pragmatic or applied approach, it quickly becomes apparent that there is no perfect ethical solution for many of the problems and challenges that cybersecurity software and data engineers may face today. Sometimes we will identify multiple competing ethical goals that must be met in making a single decision. In satisfying one goal, programmers may find that they are less successful in satisfying an equally valid, competing goal. It may become necessary to sacrifice one's commitment to one value (or set of values) to meet a different value or set of values.

Christensen et al. (2021) present a scenario where an elderly patient requires a medical operation in which complex, computer-assisted technology would be placed in their body to monitor and adjust their heart rate or dispense insulin. The data in such a machine would need to be secure to prevent the individual from being a victim of either medical hacking or simple device breakdown. They describe a "most secure solution" that might require the patient to undergo repeated operations to upgrade and monitor the technology. In contrast, they present less secure devices that spare the patient from having multiple frequent operations to upgrade and monitor the device. In this case, the competing goals are having the most secure device and data versus prioritizing patient comfort and healing.

Table 1.1 Cybersecurity Scenarios Which Present Competing Goals

GOAL 1	*GOAL 2*	*Example*
Medical device security	Patient comfort	Insertion of a pacemaker or other device
Individual privacy rights	Collective health security	Use of a COVID tracking app
	Collective road security	Training a dataset for self-driving cars
Safeguarding proprietary company data	Keeping data accessible for visually impaired employees and staff	Deciding to use sonographic charts, even if they must be forwarded to an outside provider to be created
Defeating terrorism	Treating US citizens as "innocent until proven guilty"	Use of profiling and predictive behavioral analysis

Table 1.1 describes a variety of scenarios in which designers might be asked to consider multiple ethical goals or values in designing a product.

As this example shows, what is best for the individual might not always be what is best for the organization or society as a whole. Some security solutions may also impact different groups of users in different ways. (For example, women, who are more likely than men to be stalked, may have different needs and desires in thinking about installing an in-home security solution like a camera or recording device. Women users might oppose installing a security camera in their home, since they are wary of being observed. Male users, however, might be more willing to accept a security camera since their major security worry might be the security of their home and possessions rather than concerns related to bodily autonomy and privacy.)

In resolving some situations with competing values, we need to consider the issue of trust. One's previous life experiences can impact how safe we feel sharing our private user information with others in our organizations and societies. In the same way, people may have different risk orientations. Some individuals may be willing to give up more individual autonomy to get a greater feeling of security in society, while others may not. For example, people who have grown up under repressive regimes might prioritize the value of preserving individual autonomy and privacy more highly than those who have not had such an experience.

Context is thus essential in applying ethical principles. The context or environment may determine which principle is most important to a group of people in a particular place at a particular time. Here we can consider the example of the Ring doorbell or the Blink security camera. These are devices that individuals can purchase and set up to monitor the security of their homes. In some US communities, individual Ring or Blink owners have been asked to make the data they gather (including footage of events in their neighborhoods) available automatically to their local police or sheriff's office. Such information might be used to investigate a neighborhood break-in or auto theft. However, an individual's likelihood to share information with the police force may vary greatly depending on socioeconomic status, race, and ethnicity. Those who have had primarily good encounters with the law enforcement community may be more likely to "opt in" to the programming of information sharing with law enforcement. In contrast, those with an ambivalent or unfavorable view of law enforcement may be less likely to do so.

A programmer or device designer may thus wish to consider whether these devices should automatically be set to share data with others outside the home (automatically "opting in" to data sharing to meet the goal of preserving neighborhood safety) or whether the individual should be the one to change his settings from "opt out" to "opt in" should they choose to share this data. This second solution would prioritize individual data privacy over preserving neighborhood safety through sharing and aggregating data. Choosing what's best might thus depend on the user profile and people's historical experiences, their trust in authorities, and their comfort level with taking risks.

But how should a computer programmer or engineer respond to these dilemmas? In economics, the term *satisficing* describes how consumers may sometimes choose a product that meets most of their needs but not all. (In some instances, we can use the words "least bad option" or "*Goldilocks solution*" to describe a middle-ground solution that is not perfect but also not terrible.) In the field of information privacy, for example, we can point to middle-ground solutions which can satisfy both conditions described earlier: preserving as much of an individual's privacy as possible while still simultaneously collecting, utilizing, and even storing data that can contribute to the creation of better outcomes and future outcomes. In this instance, data engineers may decide to collect individual user data and utilize various mechanisms like data anonymization or data pseudonymization so that the data is less likely to be linked back to an individual user.

Ideally, programmers would know precisely what ethical values to prioritize from the earliest stages of the design process. That way, they can design for privacy, accessibility, or security design – considering each design decision they make in all stages of the design process through the lens of this particular issue. However, it is sometimes difficult to predict the end product when technology evolves rapidly. It is also difficult to predict precisely how individuals and groups might decide to use the product once it has been created.

Here we can consider the Facebook or Meta app, initially developed by an undergraduate student interested in expanding his dating options. Should an undergraduate Mark Zuckerberg have anticipated that someday governments might wish to use the platform to collect data on individual users and their preferences to target advertisements to them during an election season? How might the platform have evolved differently and look different today if Zuckerberg's highest priority had been safeguarding information security rather than designing a platform that prioritized user openness, information sharing, and transparency? Today, his corporation is attempting to adjust the platform to treat the information gathered differently, emphasizing user privacy. New protocols and procedures are also evolving to determine who may access user information and under what circumstances. This story might thus serve as an example of how an organization's ethical priorities might evolve and change, and how a technology thus does not remain static but must also evolve to satisfy new and different ethical priorities.

The Relationship Between Ethics and Religion

As you may have noticed in considering Plato's thought, many ethical questions overlap with religious and legal questions. Ethical arguments establish standards for behavior and practices and provide the grounding for describing what constitutes a breach of ethics by an individual or group. They set expectations for behavior and describe the conditions under which people can be held accountable for violating those expectations.

Ethics requires *accountability*. Thomas Aquinas, a Christian scholar who wrote and taught in the Middle Ages, asked questions about the morality or ethics of warfare. He asked whether moral people could be soldiers and whether fighting to defend someone else was inherently moral. In Aquinas' mind, he was accountable to God; he believed that religion provided the basis for his values and that God would ultimately judge him for whether or not he had acted morally in his lifetime. For those who are religious, accountability may be to their god.

For those who are not religious, accountability is often to society. We can identify ethical schools of thought associated with the world's major religions (including Buddhism, Confucianism, Judaism, Islam, and Christianity). Still, we can also identify ethics based on the values of a community, including those of a profession. Within the environmentalist community, we encounter ethics of deep ecology. This belief system advocates for the rights of animals and living creatures like trees as equal to that of humans. People involved in international development often reference care ethics, which considers the responsibilities that one group of individuals or states may have to others, including the idea that wealthy nations should demonstrate concern regarding poorer nations and work for justice and equity. For military members, military ethics arguments – emphasizing military values like duty, honor, and country – may resonate and help inform their thinking and practices as military members.

Luciano Floridi's *information ethics* considers people's obligations regarding how they treat information – whether they engage in practices like deception or censorship, and hoarding information or sharing it. His work has many implications for cybersecurity (Floridi, 2013). Finally, we can see professional codes of ethics like the ACM (Association of Computer Machinery) and the IEEE code of ethics for software engineers as sources of moral values. These two professional codes clarify that information specialists are accountable to the public, seeking to serve in their work.

The Relationship Between Ethics and Law

You might wonder why ethics matter when we have laws. Many ethical queries could be solved simply by stating that certain practices are wrong because they are illegal or criminal without delving further into their philosophy. As long as I obey the laws, you might think, why should I think about ethics?

Indeed, some analysts follow this view in thinking about cybersecurity. Levy describes cyber ethics violations as acts that depart from the norms of a given workplace and suggests that better socialization into workplace norms can effectively solve the problem of cyber ethics violations (Levy, 1984). Similarly, Chatterjee et al. define unethical information technology use as:

> the willful violation – by any individual, group, or organization – of privacy and property and access and accuracy – concerning information/information goods resident within or part of an information system, owned/controlled by any other individual group of the organization.

They describe an act as unethical if it breaks the rules or causes harm to others – even if the individual who carried out the actions feels that they were acting ethically or in line with their convictions. That is, they argue that rule-breakers always act unethically. They believe that an outside observer can decide whether someone has acted ethically by comparing their behavior to an objective standard like an organization's professional code of conduct (Chatterjee et al., 2015).

But there are good reasons we need to learn about ethics, even when laws exist covering the same questions.

First, some analysts argue that ethics precede laws or are often formulated based on existing ethical thinking. Society's codes of law traditionally have rested on foundational understandings about what constitutes a moral action, individuals' obligation to others, what actions society should regulate, and to what end. We can look back to the first written legal code, the Code of Hammurabi, written down in Ancient Mesopotamia almost 2000 years ago, to see that societies had understandings of what constituted justice, fairness, and acceptable retribution, which then became codified into laws.

Today, Koller (2014, p. 157) refers to **conventional morality**, or "those laws or rules which hold sway in a group, society or culture because they are acknowledged by a vast majority as the supreme standards of conduct," as the basis for ethical behavior in a society. For example, conventional morality sets the expectation that parents should care for their children and even make personal sacrifices to provide for them. Or we might expect that a doctor will care for patients – even in situations where he may not be formally on duty, like at a roadside accident, and even when he might not be paid for his services. Most people believe that doctors should ensure health in society and have more outstanding obligations. (For example, we expect them to take personal risks carrying contagious patients during an epidemic.)

Such moral understandings are often codified into laws. Here, we can refer to *normative validation*, or the idea that people are more likely to conceptualize an activity as morally wrong if it is also legally wrong (D'Arcy, 2012, p. 1100). Thus, we are not surprised if parents who fail to care for children or doctors who fail to care for patients are the subject of legal proceedings – as they are seen to have violated both conventional morality and the law.

But while MacDonald sees laws as reflecting or codifying a preexisting ethical consensus, other thinkers have suggested that laws and codes of behavior provide the basis for establishing trust between citizens in a community, and that it is only in a stable, established community where people trust one another that ethics can then develop (MacDonald, 2011). The German theorist Immanuel Kant espoused the view that law necessarily preceded the establishment of ethics, since law provided the necessary constraints on human action that made ethical behavior possible.

This "chicken or the egg problem" (whether law precedes ethics or ethics precedes laws) appears in current debates about the foundation of ethics in cyberspace – a place where matters of laws, codes of conduct, and even legal jurisdictions have not yet been settled. Some analysts believe norms, values, and legal understandings will emerge organically in cyberspace. At some point, everyone will be convinced of the rightness of specific values and norms in cyberspace, which will be regarded as universal. Myskja argues that solid legal regimes need to be set to spell out the rights, responsibilities, and constraints on human behavior in cyberspace. Once that is done, he argues, ethical understandings will naturally emerge (Myskja, 2008).

But others believe that individuals, corporations, and even nations must intervene to build this consensus since it will not emerge organically. Indeed, some believe that given cyberspace's global and multiethnic nature, a consensus on values may never emerge. Luciano Floridi (2013), in his seminal work on the ethics of information, calls for creating ethical norms and understandings in cyberspace while technology is still advancing, so that the two develop in conjunction with one another. He refers to this process as "building the raft while swimming," proposing information ethics that engages with current moral and judicial understandings and anticipates and responds to problems that might arise later. Similarly, Brey argues that while ethical principles may inform legislation, laws alone are insufficient to establish or substitute for morality. Individuals still need to think of themselves as moral decision-makers and to consider their actions and how they will be perceived and affect others (Brey, 2007).

Next, laws, morality, and ethics do not always line up neatly in the real world and cyberspace. Koller (2014) speaks of *moral standards* – which differ from laws in that individuals can decide whether or not to conform to them, whereas laws are seen as having sway over everyone who resides in a region or is a citizen of the region. He notes that moral standards are often seen as having greater force than law and prioritizing laws and social customs when there is a conflict or contradiction.

Throughout history, individuals and groups have opposed laws which they regarded as unjust or unethical. Mahatma Gandhi opposed British imperialism in India and fought for India's independence, while Martin Luther King opposed discriminatory racial laws in the United States. Each man felt an ethical duty to oppose unjust laws. Here we see that ethics is related to religion and law but does not always neatly parallel. Ethical arguments may contradict both faith systems and laws – and frequently do.

In considering cybersecurity, many activists who have engaged in DDoS attacks against corporations, terrorist groups, or authoritarian governments regard their activity as a form of civil disobedience they are performing against unjust laws or decisions. For example, we can consider the actions of the online international vigilante group Anonymous, which took down Russian government websites and state-owned media outlets during the February 2022 invasion of Ukraine. Using DDoS attacks, doxing, and vandalism, the group aided Ukraine's armed forces in defending Ukraine. Many of their actions were illegal and generally unethical, but many observers believed that "the ends justify the means." They believed it is possible to carry out bad actions to achieve a good end.

Thus, cybersecurity practitioners need to be able to think ethically and critically. One difference between a profession and a mere job is that professionals often work independently, without supervision. And professionals are expected to go beyond merely reading a manual or applying a technique. They need to think critically when the rules are unclear or ambiguous, or when more than one rule applies (Gotterbarn, n.d., p. 3). They need to know where it might be okay to violate a rule and where it is necessary to draw the line and exercise restraint.

Introducing Cybersecurity Ethics

Within the field of computer ethics, many analysts ask questions related to cybersecurity. We should note here that cybersecurity has two different definitions. Social scientists, including policy analysts, often define cybersecurity as those aspects of computer security specifically related to national security issues, like cyberterrorism, and protection of national assets, like those belonging to the Department of Defense. These practitioners consider the political, economic, and social vulnerabilities created by vulnerabilities in the cybersphere. Nissenbaum (2005) describes these individuals as concerned with three types of dangers: how connectivity allows for the creation of social disruption (including the ability of hate groups to organize); the threat of attack on critical infrastructure, and threats to the information system itself through attacks on that system.

However, individuals in technology fields define cybersecurity more specifically, referring to practices and procedures used to secure data and data systems in cyberspace, regardless of who owns the systems or uses them (Brey, 2007). Here, cybersecurity protocols refer to encryption and procedures like scanning for viruses, ensuring a corporation and its employees are well-versed in cyber hygiene practices and are not vulnerable to cyber-attack vectors like phishing or social engineering. Brey further distinguishes between system security and data security. System security refers to securing hardware and software against programs like viruses. Data security or information security refers to protecting information stored in a system or transferred between systems (Brey, 2007, p. 23).

The leading professional organization associated with cybersecurity, the Association of Computing Machinery's (ACM) joint task force on cybersecurity education, defines cybersecurity as

> A computing-based discipline involving technology, people, information, and processes to enable assured operations. It involves creating, operating, analyzing, and testing secure computer systems. It is an interdisciplinary course of study, including aspects of law, policy, human factors, ethics and risk management in the context of adversaries.
>
> (Burley et al., 2017, p. 683)

In this text, we consider cybersecurity ethics. Cybersecurity ethics are *professional ethics*, providing contextualized, specific knowledge to a group of practitioners who share certain characteristics (Didier, 2010). Each profession asks, "How can I most ethically enact my profession and the project of my profession, given the constraints I face?" Those who work in cybersecurity form an *epistemic community* – they have had similar training, share a specific vocabulary and set of ideas, and belong to the same professional organizations. Over the years, many professions have developed their ethical codes, delineating a particular occupation's values and understandings. The code of ethics document lays out a list of critical organizational values. It describes the values practitioners should subscribe to and a vision for viewing their job and role in society. The code of ethics is a statement of the organization's values and often a detailed guide to practical ethics, laying out specific steps practitioners should take regarding ethical issues in various situations. We can distinguish between abstract questions like "What is justice?" and more practical ethical queries like "Should I report a professional colleague for this type of misconduct?"

But not all professional and ethical codes are alike. They may be largely aspirational – laying out standards that practitioners should regard as ideal behaviors – or they may be regulatory – laying out both standards and specific penalties that practitioners face if they do not comply with the profession's ethical code. Codes also differ in terms of their specificity. A *normative code* can spell out general principles (i.e., specifying that a lawyer, physician, or teacher should attempt to treat all clients equitably and fairly) or go into great detail about the behavior expected (i.e., stating that physicians cannot refuse patients treatment on racial, gender, national or other grounds) (Gotterbarn, n.d.). A code can thus inspire members and create a unity of purpose. Still, depending on its regulatory power and level of detail, it can also elicit member compliance to standards by requiring adherence for those wishing to enter the profession and revoking privileges and licenses for those who have violated the standards. A member may face professional censure or penalties for a legal or ethical breach in some professions. Often the two categories of breaches overlap. For example, a physician who sells a medical prescription commits both legal and ethical breaches. He would face legal charges, and the state medical board would revoke his medical license, rendering him unable to practice medicine.

Not all professions can or desire to engage in this regulatory or policing function of their member's character and ethical behavior – instead, laying out an ethical code that is aspirational or inspiring and does not have concrete measures for regulation or compliance. As a new profession evolves and becomes more professional, its code of ethics will usually move from merely aspirational to more specific and regulatory (Gotterbarn, n.d.). We can identify several ethical codes that vary in their specificity and regulatory potential in cybersecurity. The Canadian Information Processing Society (CIPS) Code of Ethics and Standards of Conduct includes filing a complaint if a member does not adhere to the code. The ACM Code of Ethics, developed in 1972 and revised in 1992, and used throughout the world, is regarded as a code of conduct as well as a code of ethics since it has three sections: canons or general principles; ethical considerations; and disciplinary rules, which involve sanctions for violators.

Because cybersecurity experts work in various settings – including hospitals, corporations, government offices, and the military – throughout this text, we also draw upon other sets of professional ethics. Cybersecurity ethics is thus an interdisciplinary practice incorporating ideas drawn from fields of study, including medical, military, legal, and media ethics. (See Box 1.2, "Going Deeper: The Hippocratic Oath," for more on medical ethics.) In addition, cybersecurity ethics often draws upon ideas from ethics of technology. The *ethics of technology* consider how new technologies can be shaped and harnessed to contribute to living a good life.

Box 1.2 Going Deeper: The Hippocratic Oath

The Hippocratic Oath is believed to have been written around 400 BC. It is named after Hippocrates, the Greek "father of medicine," but no one knows who actually wrote it. Now nearly 2500 years later, almost all doctors in the United States receive a modern version of this oath as part of their medical school graduation ceremony. The oath serves as a code of conduct and unifying statement of the ethics of the medical profession for all doctors. In it, doctors promise to "do no harm," not to fraternize with their patients, to respect their patients' privacy and confidentiality, and to treat them with respect.

Hippocratic Oath: Modern Version

I swear to fulfill, to the best of my ability and judgment, this covenant:

I will respect the hard-won scientific gains of those physicians in whose steps I walk, and gladly share such knowledge as is mine with those who are to follow.

I will apply, for the benefit of the sick, all measures [that] are required, avoiding those twin traps of overtreatment and therapeutic nihilism.

I will remember that there is art to medicine as well as science, and that warmth, sympathy, and understanding may outweigh the surgeon's knife or the chemist's drug.

I will not be ashamed to say "I know not," nor will I fail to call in my colleagues when the skills of another are needed for a patient's recovery.
I will respect the privacy of my patients, for their problems are not disclosed to me that the world may know. Most especially must I tread with care in matters of life and death. If it is given me to save a life, all thanks. But it may also be within my power to take a life; this awesome responsibility must be faced with great humbleness and awareness of my own frailty. Above all, I must not play at God.
I will remember that I do not treat a fever chart, a cancerous growth, but a sick human being, whose illness may affect the person's family and economic stability. My responsibility includes these related problems, if I am to care adequately for the sick.
I will prevent disease whenever I can, for prevention is preferable to cure.
I will remember that I remain a member of society, with special obligations to all my fellow human beings, those sound of mind and body as well as the infirm.
If I do not violate this oath, may I enjoy life and art, respected while I live and remembered with affection thereafter. May I always act so as to preserve the finest traditions of my calling and may I long experience the joy of healing those who seek my help.

Written in 1964 by Louis Lasagna, Academic Dean of the School of Medicine at Tufts University, and used in many medical schools today.

Source: Found at www.pbs.org/wgbh/nova/body/hippocratic-oath-today.html

Ethics in Policy

As we see in considering subfields like medical ethics, an ethical framework lets us ask questions about an individual's ethics, the ethics of a profession, and the ethics of a society or group of people. Ethics may refer to a set of rules about what an individual should value (as Plato believed) and how an individual should behave, but it might also refer to upholding the values and beliefs associated with one's profession or community or even one's nation.

As we will see throughout this text, sometimes ethical concerns overlap with policy concerns. An ethical stance may lead individuals to advocate for specific laws and policies. Thus, a US government official who works on refugee issues may wish to choose the right and just response but may also face questions about which solutions are politically feasible, are legal, and require substantive changes to legislation and policies.

In this textbook, we focus primarily on the first two levels of analysis – the ethics of the individual computer practitioner and the ethics of the cybersecurity profession. We will not focus on the ethics of the technology itself or companies or corporations like Google, Microsoft, or the Department of Homeland Security. We will not consider legislative or policy changes or dwell on the morality of specific policies, such as US national security policy in cyberspace.

Instead, beginning in Chapter 2, we introduce frameworks for asking ethical questions. We also provide you with information about your professional obligations as a member of the cybersecurity profession. However, in the next section, we look at the larger question of the ethics of technology to provide you with the necessary background to move forward in thinking about your ethics in this field.

The History of Computer Ethics

Computer ethics is an interdisciplinary field. Norbert Weiner was the first computer scientist to pose ethical questions about the field. Weiner, a mathematics professor at the Massachusetts Institute of Technology in the 1950s, asked whether the information was matter or energy. He suggested that computing technologies differed in fundamental ways from other technologies and spoke about a future of "ubiquitous computing" in which we would live in a world surrounded by computers that both collected data about us and provided us with data continuously (in this way anticipating what we now refer to as the Internet of Things). Weiner's work, considered very revolutionary at the time, provided a basis for studies today in fields as diverse as library, journalism, and medical ethics.

Other philosophers added to the debate later, shaping it in new directions. In 1976, Walter Maner posed questions about technology and its use in medicine. He argued that developments in computing, like connectivity, had created unique ethical problems that had not existed before. His colleague at Old Dominion University, Deborah Johnson, disagreed. In her work, she argued that computers might give preexisting ethical problems "a new twist" but that existing philosophical ideas could be shaped and applied to these new and emerging issues. Johnson went on to write the first textbook in computer ethics (Bynum, 2015).

In 1976, Joseph Weizenbaum, an MIT computer scientist, published the landmark book *Computer Power and Human Reason*. He had developed a computer program called ELIZA, which simulated a person's interactions in conversing with a psychotherapist. He began thinking about how humans interact with computers and the ethical issues raised after observing the attachments his students and others developed to ELIZA (Bynum, 2000).

James Moor (1985) moved the "*uniqueness debate*" in an article in *Metaphilosophy*. He argued that computers allowed humans to go beyond their previous human abilities to do things that previously would have been impossible (i.e., performing mathematical calculations at superhuman speed). As a result, he argued, a computer specialist might find themself in a situation where a *novel problem* or *policy vacuum* emerges. There is not yet any clear law or even ethical framework (Moor, 1985). For example, in recent years, developers have created websites for those who want encouragement to adopt an anorectic lifestyle, for individuals who wish to receive support from others before committing suicide, or for the sale or resale of illegal drugs. In each of these situations, the crime was so new that it was unclear what existing legal statutes applied. Today, novel problems exist regarding ownership rights of materials held in cloud storage; data is created through the Internet of Things, and monetary issues arise from digital currencies like Bitcoin. Scholars and analysts are also attempting to formulate rules regarding whether an individual who has a warrant to collect data also has the right to store that data, compile that data into some form of an aggregate file – with or without removing identifying information, or pass on that data including selling it, sharing it, or giving it to the government.

In situations like technology lags, grey areas, or unjust laws, a coder or security professional needs to hone their own ability to think ethically, act independently, or even guide others within their organization or corporation. (Note here that when we address these grey areas in this course, if laws are addressed, the assumption is that the student is an American student in an American classroom. Most references are to US law – for example, copyright – and the assumption is that the norms prevalent in America regarding these issues are the default. We are also not necessarily advocating that readers take particular actions – including those that would violate any type of laws or rules within their organizations, or locally, on the state or federal levels.)

Box 1.3 Critical Issues: Plagiarism, Paraphrasing, and Artificial Intelligence

What does it mean to be an author? Standard definitions tell us that an author is someone who writes a text. However, as technology advances, it is becoming harder and harder to answer what it means to author something.

You may have been asked to write essays and research papers throughout your academic studies. Much of that task undoubtedly required you to read an article and then paraphrase or provide a gist of the author's main arguments. In doing so, you were able to show your professors that you had read an assigned text, had engaged with the text sufficiently to understand its meaning, and were thus capable of putting the main ideas into your own words. Writing a paper provides your professors with a gauge of whether you understand the material you encounter and whether you are doing sufficient work in your classes (Rogerson, 2017).

However, new linguistics and artificial intelligence developments now make it possible to "outsource" many writing tasks to an algorithm. Today, paraphrasing software can alter sentences, paragraphs, articles, and books. Sites like Quillbot and Wordtune allow users to upload blocks of text (often for free) and receive an altered text back. Such programs carry out tasks including finding synonyms, altering the arrangement of words and sentences, and abbreviating and combining sentences. More sophisticated programs exist that can even "spin" an entire article – carrying out a search for information online and then paraphrasing and rearranging the information into a report or essay.

The new text can outwit plagiarism checking software like Turnitin or Safe Assign because the original text is altered just enough that it does not cause an alert to the plagiarism checking algorithm. In submitting an altered text, you might fool your professors. They may think that you have done the reading and have understood it well enough to paraphrase it.

So is using paraphrasing software wrong, and if so, why? We can identify four ethical issues related to the use of paraphrasing software. First, if the software is used to "spin" content on a site that people access, there is a risk that they could potentially receive incorrect or potentially harmful information. (For example, an algorithmically generated article about cancer treatments might contain falsehoods and inaccuracies.) Thus, the software has the potential to harm consumers of the information it generates.

Next, submitting work that is not one's own originally authored work represents a form of deception and a violation of a university's academic integrity principles. It also constitutes a form of **plagiarism**, since it rests on taking credit for the work without attribution.

Finally, in some instances, paraphrasing software can be used to commit theft if, for example, one uses copyrighted content (like a novel or play) on one's site, presenting it as one's work, and profiting from doing so. Contract cheating refers to ethical violations where a student buys a paper online. In such instances, paraphrasing software might be used to "write" the original essay. An expert then adds a human touch by fixing any issues that might arise with the machine-generated text – profiting from someone else's work (Prentice & Kinden, 2018).

However, some ethicists today note that our definitions of knowledge and how we consider knowledge generation may be gradually changing. If, for example, an academic writer uses a program like Grammarly to improve their writing – and in some instances, Grammarly rewrites an entire sentence or paragraph – does the author err in taking credit for the bot's work in this instance? Could the piece be said to have been co-authored by the writer and Grammarly? And if so, why is this co-authoring allowable while the other is not? Here we can consider the intent of one's decision to use paraphrasing software. Since the academic is not intending to deceive an audience, misrepresent the depth of one's knowledge, or profit by receiving a grade, such use may be regarded as ethical. Furthermore, standards in the field seem to point towards allowing such use of the software.

Finally, some ethicists note that the ethical issues generated by paraphrasing software are not only those that the students or writers themselves encounter. Instead, they argue that universities and academic journals are responsible for ensuring that standards are high and that work is correctly vetted so that plagiarized or stolen work is not published (Petrisor, 2021).

This short discussion of plagiarism shows how technology interacts with ethics today at a level familiar to most students. It also shows how thinking about ethics can change as technology develops. Today, ethicists are keenly involved in debates about what it means to author something, what it means to know something and how humans should and should not depend on algorithms in their quest for knowledge.

Bibliography

Petrisor, A. I. (2021). Predation, plagiarism and perfidy. *Portal*. Retrieved from https://eric.ed.gov/?id=EJ1317782.

Prentice, F., & Kinden, C. (2018). Paraphrasing tools, language translation tools and plagiarism: An exploratory study. *International Journal for Educational Integrity*, *14*(11), https://doi.org/10.1007/s40979-018-0036-7

Rogerson, A. M. (2017). Using internet based paraphrasing tools: Original work, patchwriting or facilitated plagiarism? *International Journal for Educational Integrity*, *13*(2), 1–15.

Many professional organizations and publications consider information and professional ethics, which must evolve to accompany thinking in this field. We can point to Donald Gotterbarn's Software Engineering Ethics Research Institute (SEERI) at East Tennessee State University and the annual ETHICOMP conference on Ethical Computing. Gotterbarn's emphasis on the development of professional ethics is reflected in creating the

ACM's Code of Ethics and Professional Conduct in 1992, as well as in creating licensing standards through the Computer Society of the IEEE (Institute of Electrical and Electronics Engineers). Many of these initiatives were international in scope, reflecting the understanding that computer ethics is universal and not specific to a particular nation, belief system, or type of educational institution (Gotterbarn, n.d.).

Information ethics is a diverse and interdisciplinary field in which scholars from sociology, anthropology, psychology, engineering and computer science, business, and the law come together to consider how individuals think about computers, how they interact with them, and what their relationships are with computers – even what their emotions are (Sellen et al., 2009, p. 60). Libraries attempt to clarify the credibility of internet sources, while those involved in local, state, or national government discuss equitable access to internet resources to questions about privacy and secrecy. Information ethics can also include media ethics and, increasingly, the ethics of information management (i.e., how corporations and government agencies treat your data). Ethical thinking regarding whether and under what circumstances information generated or stored should be regarded as personal effects of medical practitioners, teachers and professors, lawyers, and others. It may be the subject of government, associations, and policymaking.

Why Cybersecurity Ethics Matters

Computer scientists can impact the environment through the internet. As a result, they affect events online and in the real world. Throughout this book, internet technology provides the foundation for other institutions, allowing people to contact their legislators and government, educate themselves, and access other social and political institutions like housing offices and emergency services. Internet technology also provides the foundation for other rights – making it possible for people to engage in freedom of assembly, religion, and speech. A free and open internet is thus part of a free and open society. Decisions individuals make about the design and functioning of that system have global, powerful, and permanent ramifications.

Thus, it is essential to understand computer ethics and the underlying principles. You cannot merely obey the laws or go with the flow – since every practitioner will be asked at some point to decide independently about which files to keep, where materials should be stored, who can access them, and what credentials will be required to access materials and so forth. There are not merely technical issues. They are also ethical issues.

CHAPTER SUMMARY

- Ethics is a branch of philosophy concerned with "the good."
- *Normative ethics* considers what one ought to do, while descriptive ethics is concerned with observing and understanding ethical behaviors in different places, times, and cultures.
- Computer ethics are normative ethics. They are a branch of both practical and professional ethics.
- Moral philosophers can be either *moral relativists* or *objectivists*. Moral relativists argue that what is morally right or wrong can depend on someone's disposition or the conventions of a particular historical era or culture. Objectivists believe that it is possible

to find out what the right thing to do is objectively and that values and value commitments are universal in scope.

- In some instances, it may be necessary to choose among competing values. People may rank order the values they wish to prioritize differently, depending on factors like culture and their historical experiences.

DISCUSSION QUESTIONS

1 **AI and morality**
Watch the TED Talk by Zeynep Tufekci, "Machine Intelligence Makes Human Morals More Important." Found at: www.ted.com/talks/zeynep_tufekci_machine_intelligence_makes_human_morals_more_important

- Summarize Tufekci's argument.
- Why does she feel that humans will ultimately be the arbiters of morality?
- What does she mean when she says machine learning is opaque?
- Does the person who develops an algorithm for hiring that includes problems like those described by Tufekci (such as screening out candidates who might someday develop depression) owe anything to the human resources manager?
- Who would you regard as ethically or morally accountable in a situation like she describes – the computer, the programmer, the algorithm designer, or the human resources manager?

2 **Considering your own experiences**
Consider your own experiences in the area of computer science. Can you think of any ethical dilemmas you have encountered in your work? What were they, and how did you resolve them?

3 **Finding a consensus**
What are some issues you have encountered in your work or leisure online for which there is no consensus regarding the right thing to do?

RECOMMENDED RESOURCES

Christensen, M., Gordijn, B., & Loi, M. (2021). Introduction. In M. Christensen, B. Gordijn, & M. Loi (Eds.), *The ethics of cybersecurity* (pp. 1–11). New York: Springer Open.

Floridi, L. (2013). *The ethics of information*. Oxford: Oxford University Press.

Nissenbaum, H. (2005). Where computer security meets national security. *Ethics and Information Technology*, 7(2), 61–75.

References

Audi, R. (1995). *The Cambridge dictionary of philosophy* (2nd ed.). New York: Cambridge University Press.

Brey, P. (2007). Ethical aspects of information security and privacy. In N. A. Petkovic (Ed.), *Security, privacy and trust in modern data management* (pp. 21–36). Heidelberg, Berlin: Springer.

Burley, D., Bishop, M., Kaza, S., & Gibson, D. H. (2017). *ACM joint task force of cybersecurity education*. Seattle, WA: ACM.

Bynum, T. (2000). *A very short history of computer ethics*. Newark, DE: American Philosophical Association Newsletter on Philosophy and Computing.

Bynum, T. (2015). Computer and information ethics. In E. N. Zalta (Ed.), *Stanford encyclopedia of philosophy (Summer 2020 Edition)*. Retrieved from https://plato.stanford.edu/archives/sum2020/entries/ethics-computer/.

Calman, K. (2004). Evolutionary ethics: Can values change? *Journal of Medical Ethics*, *30*(3), 366–370.

Chang, C. L.-H. (2012). How to build an appropriate information ethics code for enterprises in Chinese cultural society. *Computers in Human Behavior*, *28*(2), 420–433.

Chasi, C. (2014). Ubuntu and freedom of expression. *Ethics and Behavior*, *30*(2), 495–509.

Chatterjee, J. S., Sarker, S., & Valacich, J. (2015). The behavioral roots of information systems security: Exploring key factors related to unethical IT use. *Journal of Management Information Systems*, *31*(4), 49–87.

Christensen, M., Gordijn, B., & Loi, M. (2021). Introduction. In M. Christensen, B. Gordijn, & M. E. Loi (Eds.), *The ethics of cybersecurity* (pp. 1–11). New York: Springer Open.

D'Arcy, J., & Devaraj, S. (2012). Employee misuse of information technology resources: Testing in a contemporary deterrence model. *Decision Sciences*, *43*(6), 1091–1124.

Didier, C. (2010). Professional ethics without a profession: A French view on engineering ethics. In I. A. van de Poel (Ed.), *Philosophy and engineering: Philosophy of engineering and technology* (pp. 161–173). Berlin: Springer Science+Business Media B.V.

Floridi, L. (2013). *The ethics of information*. Oxford: Oxford University Press.

Koller, P. (2014). On the nature of norms. *Ratio Juris*, *27*(2), 155–175.

Levy, S. (1984). *Hackers: Heroes of the computer revolution*. New York: Penguin Group.

MacDonald, E. (2011). *International law and ethics: After the critical challenge – Framing the legal within the post-foundational*. Boston, MA: Brill.

Moor, J. (1985). What is computer ethics? *Metaphilosophy*, *16*(4), 266–275.

Myskja, B. K. (2008). The categorical imperative and the ethics of trust. *Ethics and Information Technology*, *10*(1), 213–220.

Nissenbaum, H. (2005). Where computer security meets national security. *Ethics and Information Technology*, 7(2), 61–75.

Sellen, A., Rogers, Y., & Harper, R., & Rodden, T. (2009). Reflecting human values in the digital age. *Communications of the ACM*, 58–66.

2 Three Ethical Frameworks

LEARNING OBJECTIVES

At the end of this chapter, students will be able to:

1. Articulate assumptions of the three frameworks – virtue ethics, practical ethics, and communitarian ethics
2. Compare and contrast major ethical stances, including virtue, utilitarian, and deontological
3. List criticisms of each of the three ethical lenses
4. Apply the three different ethical stances in thinking through the ethical consequences of a particular problem or action

This chapter examines three ethical stances for considering ethical problems – virtue ethics, utilitarian ethics, and communitarian ethics.

Why Use an Ethical Framework?

Every day we make hundreds of decisions. Should I walk to work, drive, or take public transportation? Should I bring my lunch from home or purchase it? Should I apply for that promotion or decide to be happy with my current job?

Academics in many fields – including business, leadership, economics, and psychology – think about how people make decisions. In doing so, they build *models* – or simplified pictures of reality – to understand decision-making. In building a model, the researcher specifies her assumptions about how it works, the conditions under which it works, and any restrictions or constraints that might affect its working. In considering decision-making, we as researchers make several assumptions. First, we assume we can identify an individual or group as the decision-maker in any situation (that is, we assume that decisions don't just evolve in an organization and are not made by consensus). Second, we assume that decision-makers are aware they are deciding and have the authority to make decisions. Third, we also assume that they have agency or control over the choice of actions being considered and that they are not being coerced to choose one outcome over another. Fourth, we assume that decisions are made in isolation, that one decision does not necessarily affect and is not affected by any previous decisions, and that the decider is aware of the constraints under which they are making the decision. (For example, if one considers choosing to violate company policy, we assume that the individual knows the policies and consequences of choosing to violate them.)

DOI:10.4324/9781003248828-3

In this text, we apply three models of decision-making – virtue ethics, utilitarian ethics, and deontological ethics – to ask questions about cybersecurity. Should you send spam e-mails, and what is the consequence? Should you engage in cyberwarfare against your own country on behalf of another country paying for your expertise? How should you treat potentially embarrassing or incriminating information you find about an employee while monitoring a computer system? In each case, we assume that the individual is the decider, that he is aware of his position as a decider, and that his decision is independent of previous and subsequent decisions.

Depending on the chosen model, the lenses emphasize different concepts – such as identity and utility. Sometimes the model's recommendations will contradict one another, while in other instances they might overlap. The models can disagree about the badness or goodness of a proposed action or attitude, the moral or ethical significance of a proposed action, or why it matters. Sometimes all three lenses might yield the result that an action – like pirating copyrighted content and redistributing it to make a profit – is wrong, but they may disagree about why it is wrong. In other instances, all three models might suggest taking a specific action, but they may disagree about the why or what makes something necessary. For example, both virtue ethics and deontological ethics might value the achievement of equity and nondiscrimination in a situation. Still, they might disagree about why equity is necessary and what choosing equity means. In brief, a virtue ethics argument rests on a claim that "I am the kind of person who," while a deontological ethics argument considers those who might be the subjects of any formulated ethical rule.

In addition to using the models, philosophers use *thought experiments*. Ethical philosophers often tell a story about a situation in deciding what the correct decision is in a particular situation. The situation may be outlandish, far-fetched, or unlikely (i.e., Jim is informed that he will be required to kill one of his family members this evening at seven o'clock). Such a hypothetical situation is unlikely to happen in real life. However, the story can then explore the motives and consequences of an ethical choice.

Computer ethicists may use stories drawn from science fiction to allow individuals to think through the consequences of emerging technologies. For example, in 2000, law professor Lawrence Lessig (2000, p. 99) posed a thought experiment on his blog, asking, "Could architects ever build a building which was so safe and secure that no one could ever break into it?" He then explored the consequences for society, criminality, and law enforcement of the existence of such a building. More than ten years later, we saw these concerns reflected in discussions about whether the Federal Bureau of Investigation could order Apple Corporation to decrypt or unlock users' cell phones if they were implicated in terrorist events, like those in San Bernardino in fall 2015.

Philosophers in many fields of applied ethics use a variation of what has come to be known as the *Trolley Problem*, developed in 1978 by Philippa Foot. In this thought experiment, children play on the trolley tracks, unaware that a trolley is speeding toward them. An observer stands on a bridge. From this vantage point, he sees the events as they are about to occur and can influence them. He stands next to a huge man. This man's mass is significant enough for the observer to throw him down onto the tracks; his body could stop the trolley. The man would, however, die. Thus, the observer needs to decide if it is moral to sacrifice one man's life to save the lives of the children on the tracks. Today, ethicists point out that automobile drivers may sometimes be called upon to make a similar decision. If they are about to hit an oncoming car, should they swerve to avoid it, saving the life of that car's occupants but possibly endangering the occupants of their vehicle? To whom is their ethical duty more excellent – that of the stranger or their own family or friends? Robot ethicists note that we will soon have self-driving vehicles and ask how computers can learn to think about similar ethical scenarios when they react to unanticipated situations and road obstacles (Jaipuria, 2015).

However, some academics have criticized philosophers' reliance on models and thought experiments. Walsh (2011) notes that neither models nor thought experiments reflect the real world, where we seldom act in isolation and our choices are often not discrete or made in isolation. We often do not conceptualize ourselves as being at a decision point when we act.

Nonetheless, we rely upon these models for both pedagogical and research reasons. The three lenses provide students with a straightforward way of thinking about ethical decision-making and thinking through ethical decisions. And, as you will see in Chapter 10, these same models are often used by policymakers, including those in the US military, as they think through the consequences of ethical decision-making about issues including the use of autonomous weapons and the rights and responsibilities of soldiers engaged in cyberwar.

Box 2.1 Going Deeper: Three Paradigms for Thinking About Ethics

As we think about what makes a technology ethical, we may ask the question in three different ways. Each paradigm or lens allows us to ask different ethical questions about technology by emphasizing different ethical issues.

Values-based ethics allows us to think about technology's values, and technology either supports or opposes them. For example, in thinking about technologies that allow (or require) users to download an app that gives a company access to information about their behavior, values-based ethicists will likely reflect on how such technology can contribute to or degrade user privacy in the abstract. A values-based ethicist might thus ask how having access to technology like Alexa might change how people think about their home as a private space. If we allow electronic eavesdropping, they might ask, will we eventually lessen the distinction between public and private spaces? They might ask questions about values like equity – asking if some people might be more likely to "trade" their privacy for another good the app is offering. (For example, would people who struggle financially be more likely to install a free video or music app – even if it tracked what music they listen to or what videos they watch – than someone who does not struggle financially?)

In contrast, someone who uses a design-based paradigm considers how to build an ethical technology and the sorts of decisions that a builder must make. In considering a human–machine interface that collects information about its users, design-based paradigms would focus on questions like: Does a company owe it to users to let them know that they are being tracked by an app they download? Does the app have an ethical requirement to allow people to opt out of tracking their behavior by changing the settings on their device? They might also ask how one app is integrated with other apps – asking if data will be shared between apps and whether people have to be informed that their data will be shared between apps. Those who consider design-based ethics might also consider what sorts of technologies could be developed which can help shield user privacy through practices like anonymization and semantic privacy (or "cloaking").

Design-based paradigms also allow us to consider **security by design**, **privacy by design**, and **accessibility by design**. Such practices allow us to ask: who is most likely to be excluded from using new technology, and who is most likely to be included? What can be done to make technology more accessible to all, and what sorts of modifications might be necessary?

Finally, application-based ethics suggests that the most critical question to ask when developing new technology is: How will the tool be used, and is there a potential that it will be used in a way that causes harm? Here, some ethicists find it helpful to ask: What if someone like Hitler had access to this technology? Is there a way that this seemingly harmless and fun technology might be weaponized or deployed against its users? Ethicists working in this vein might raise concerns about how a bad actor could use a device like Alexa to collect information about people, with the eventual aim of taking away their rights.

As this short discussion shows, the three paradigms or lenses – values-based ethics, design-based ethics, and application-based ethics – can overlap in some ways while raising distinct questions within each sphere. Therefore, it is theoretically possible for new technology to pass the values-based ethics test. However, it still can cause concerns about using it (application-based ethics). For that reason, all three ethical paradigms can be used to raise questions in evaluating new and emerging technologies.

What Are Virtue Ethics?

The first model is **virtue ethics**. We can trace this approach to ethical thinking back to the days of Ancient Greece and, in particular, to *Aristotle* (384–322 BC). Aristotle believed that everything that exists in nature does so for a purpose. This includes man, whose purpose in life is to act well as a human being (Afthanassoulis, 2012, p. 55). If he does this, Aristotle counsels, he will have a good life. This principle has sometimes been referred to as the cultivation of human flourishing. Aristotle himself describes acting in this way as the equivalent of being healthy. One is ethically healthy if one cultivates certain attitudes, characteristics, and ways of being, then expressed appropriately (Gottlieb, 2009, p. 21). Koller (2014, p. 32) writes that:

> The concept of virtue refers to the character traits of persons, their practical attitudes or dispositions, which have some motivating force for their conduct.

In Aristotle's ethics, the acts matter and the attitudes that lead individuals to carry out these acts. It is called *agent-centered ethics* since it emphasizes the decision-maker's agency or free will to make choices. It focuses on the decision-maker and his character as the primary consideration in determining whether an action is ethical. In Aristotle's view, an ethical individual is someone who does and feels "the right things at the right time in the right way and for the right reasons" (Gottlieb, 2009, p. 21). In his work, Aristotle counsels people to find the spot in the middle of the mean when thinking about character. Each virtue exists along a spectrum; neither excess nor an absence of quality is good. Indeed, displaying either too much or too little of something is a vice. For example, we can think of a spectrum ranging from timidity to courage to foolhardiness. Aristotle would counsel against having too much courage if it leads to making stupid decisions and against having too little. He also acknowledges that the right amount of virtue might vary depending on the context and limitations. For example, deciding not to rescue a drowning man would be wise if you were a poor swimmer but would be wrong if you were a good swimmer. The excellent swimmer is timid here, while the poor swimmer is wise (Gottlieb, 2009, p. 30). Similarly, in a culture like the United States, where taking risks is often regarded as praiseworthy, quitting your job to start your own business might be considered courageous; in contrast, it might be regarded as foolish in another culture (Walker, 2007).

Virtuous Does Not Mean Saintly

Although we use the word *virtues*, Aristotle's ethics aims to teach people how to live well with their human nature. It does not merely apply rules but considers people's psychology and emotions. It tells people how to "succeed" in life (or flourish) by acting sensibly and keeping with their human nature. It neither counsels taking unnecessary risks nor suggests that you must always sacrifice yourself for others – though it suggests that one should strive to treat others well if possible. Aristotle does not say you should never get angry (as a saint might) but says you should be appropriately angry at the right time for the right reasons (Gottlieb, 2009, p. 21).

Many of the virtues which Aristotle addresses overlap with the character traits (or theological virtues) stressed in many of the world's major religions. The Ancient Greeks acknowledged Four Cardinal Virtues – prudence, courage, moderation, and justice (Gottlieb, 2009). Other analysts have added character traits that individuals should develop: reasonableness, truthfulness, honesty or sincerity, goodness or benevolence, helpfulness, friendliness, generosity, humility, and modesty (Koller, 2014).

Ali (2014, p. 10) identifies virtue ethics (and practical ethics) in Islamic thought. He notes that both the Koran and the Hadith stress the importance of intent. He notes that the Prophet Muhammed stated that "God examines your intentions and actions," arguing that honest or bad intention determines a work's outcome. He also identifies ten Koranic virtues – "forbearance, generosity, adherence to accepted custom, righteousness, patience, thankfulness, flexibility, reason, sound faith, and knowledge." Jewish scholars point to the emphasis throughout the Old Testament on Wisdom. A wise person also has the tools to learn about and acquire other virtues and solve issues (Borowitz, 1999).

Virtue ethics also assumes that one's character develops over time through experiences and exposure to people and ideas. One can practice ethical behaviors and grow in virtue. In writing about the development of a virtue ethic within a military community, Valor defines virtues as "*habituated* states of a person's character that reliably dispose their holders to excel in specific contexts of action and to live well generally" (Vallor, 2013, p. 3, emphasis added).

Contemporary Virtue Ethics

Today, analysts cite two modern texts describing more recent developments in virtue ethics – Elizabeth Anscombe's 1958 essay, "Modern Moral Philosophy," and the work of Alasdair Macintyre. These texts argued that the previous philosophical emphasis on rights and duties in figuring out what people should do was insufficient and unduly focused on applying rules and finding laws for behavior. Furthermore, it focused too much on thinking about specific values like equity and justice, to the detriment of considering issues like generosity, friendship, and charity (Walker & Ivanhoe, 2007).

These theorists liked virtue ethics' emphasis on intent (or what Aristotle described as correct thinking). In Aristotelean virtue ethics, an action is considered virtuous if performed correctly, at the right place and time, and with the right intent. Here, we can consider activities that individuals can perform that seem virtuous but are not because of their intentions. In the first example, John saves a man from drowning in a lake – but only because he wants to impress the girl he is with by behaving heroically in front of her. In the second example, Kate accidentally donates to charity by pushing the wrong button on her computer during online shopping. (She meant to purchase a sweater!) In each instance, the individual did not behave morally because the individual did not behave with moral intent (Afthanassoulis, n.d.).

Therefore, virtue ethics helps apply ethical thinking since it can help us know when to do good versus merely knowing what good is in the abstract. For this reason, virtue ethics is also instrumental in framing a professional ethic and building professionalism (Walker & Ivanhoe, 2007). Indeed, we will see in Chapter 3 that many aspects of the hacker ethic stem from virtue ethics.

Critiques of Virtue Ethics

However, some do not like the virtue ethics model. We can identify three critiques: First, some scholars oppose using a model to "do ethics," arguing that models simplify the real world by definition. Here analysts point out that often in the real world, people may be coerced or compelled to make decisions through peer or social pressures or threats.

Next, some analysts suggest that the individual level is not the most appropriate level of analysis for thinking about the development of good in society. Some analysts prefer pragmatic ethics. *Pragmatic ethics* is a subfield of ethics that focuses on society rather than on lone individuals as the entity that achieves morality. John Dewey, a theorist of pragmatic ethics, argued that a moral judgment may be appropriate at one age in a given society but may cease to be appropriate as society progresses. Pragmatic ethics thus acknowledges that ethical values may be dynamic and changing.

Still, other scholars critique the specific model of Aristotelian virtue ethics rather than the idea of relying on models itself. Voicing what is commonly known as the "situationist critique," several scholars have utilized psychological research to argue that one cannot honestly speak of one's moral character or personality as a fixed and unchanging entity. These scholars note that people behave differently in different situations – even at different times of day (Kouchaki, 2014) – while virtue ethics assumes one's notion of the self is relatively stable and fixed (Slingerland, 2011). Others call Aristotelian virtue ethics western-centric since it recommends pursuing characteristics and attributes that may be prized in some societies more than others or more highly prized amongst men than women (Barriga, 2001). For example, does every society and gender equally prize courage as a value? In some societies, members might see courage as belligerence and prefer cultivating peacefulness. Finally, some scholars describe virtue ethics as based on circular reasoning: One acts in a certain way to show one's character, which is a function of one's actions. It is unclear whether actions cause character or character causes action.

Some scholars dislike virtue ethics' emphasis on individual moral reasoning, suggesting that it is selfish for the decision-maker to care more about his flourishing than the overall outcome of the decision. For example, suppose someone decided not to lie to an authority figure because he prized truthfulness. In that case, he might feel virtuous in telling a Nazi officer who hid Jewish citizens in his home amongst his neighbors. Here his virtues might lead to a lack of compassion towards others.

Virtue Ethics in Cyberspace

In Chapter 1, we encountered the uniqueness debate – the dispute about whether cyberspace is an alien environment in which traditional ways of thinking about ethics are irrelevant, or whether cyberspace is an extension of real space (or meat space), where humans behave in the same way that they do in the real world. Philosopher Bert Olivier (2013, p. 1) feels that virtue ethics apply to cyberspace. He asks:

> Why would one's virtual presence in cyber-space give you licence to behave any way morally than under the concrete conditions of the human, social life-world? In ordinary social and inter-personal interactions we expect people to behave in a morally "decent" manner; in cyberspace there should be no difference.

Indeed, studies show that individuals can import their values from fundamental world interactions into cyberspace. Harrison describes students who chose not to engage in online cyberbullying, noting that many of these students described the values which led them to make this decision. Among those values listed are caring for others, individual self-discipline or restraint, compassion, humility, and trust. As we see in this text, many interactions among individuals and groups in cyberspace rest on core values of virtue ethics, such as establishing trust, building relationships, and a desire to speak honestly and avoid deception.

Vallor (2013, p. 120) identifies 12 "techno-moral virtues" that she believes can serve individuals as a guide to making ethical and moral decisions about their conduct in cyberspace. She lists "honesty, self-control, humility, justice, courage, empathy, care, civility, flexibility, perspective, magnanimity, and wisdom." She suggests that individuals in technological fields can practice these virtues daily.

However, one might also argue that cyberspace's distinct architecture facilitates specific actions and practices – including deception. That is, the nature of the internet makes specific data collection, retrieval, storage, and sharing practices more likely and desirable. One can thus argue that someone who aggregates datasets and accidentally or on purpose learns information about an individual that may violate their privacy, and personal boundaries has made an ethical misstep by failing to respect that person. While it may violate one's values to harm a person, one can argue that harming someone's digital reputation by publishing private data is not equivalent, since people should expect that their data will likely be collected and shared by utilities like Spokeo in our current age. The one who does so is thus not a thief or person of poor moral character; instead, that person may be a businessman, analyst, or investor following standard protocols and expectations.

How Do Professional Codes of Conduct Reflect Virtue Ethics?

Virtue ethics often form the basis of a *professional code of ethics*. For example, all medical personnel subscribe to the Hippocratic Oath, which describes the responsibility a physician or health care provider should feel toward a patient. In applying virtue ethics, professionals might ask: "What does it mean to be a benevolent doctor, teacher, lawyer, or soldier? What characteristics, values, and habits should a member of my profession cultivate and represent?" We can also see virtue ethics approaches applied to psychology, social work, and nursing in human services. In each case, writers link someone's vocation or calling to practice a particular profession, their values, and their identity as a person.

What Are Deontological Ethics?

The second model is deontological ethics. We commonly point to *Immanuel Kant* as the father of deontological ethics. Kant (1724–1804) was a German philosopher who believed humans have a privileged place in the universe due to their ability to reason. Kant felt that humans could use reason to derive normative or ethical stances during the Enlightenment. Kant suggested that you could be moral even without believing in God or natural law because you could calculate the appropriate moral action or set of duties. Kantian ethics is also called deontological or ethics of *duty or obligation*.

Unlike Aristotle, Kant was not focused on the agent's state of mind or character. And unlike utilitarian ethics, Kant did not focus on the outcome of the agent's decision. Instead, deontological ethics define certain behaviors as humans' moral duties or obligations. These duties exist independently of any good or bad consequences that they might create (Brey, 2007). A deontological approach deems an action moral or ethical if the duty has been complied with. This approach does not promise that the individual who decides to fulfill his duty will necessarily be happy due to doing so, nor will it necessarily lead to the best possible outcome. Instead, it is simply the right thing to do.

Deontological ethics suggests that humans can use their reason to solve an ethical problem by searching for the **categorical imperative**, expressed as "Always act on the maxim or principle which can be universally binding, without exception, for all humans." Kantian ethics thus assumes that it is not moral to have a different set of ethical or moral laws for one individual than would be appropriate for everyone. It suggests that everyone should define and agree to adhere to the same set of standards, rather than identifying, for example, one set of standards for CEOs or wealthy nations and a different set of standards for employers or poorer nations.

The second principle of Kantian ethics is the notion of *reversibility*, or the Golden Rule. In contemplating taking action, the actor needs to ask himself, "Would I be harmed if someone took the same action against me? How might I be harmed?" Thus, he would conclude that theft is terrible since he would be harmed if someone stole from him. Trust is a fundamental component of ethics. Kantian ethics helps establish a foundation where people can trust one another – since individuals recognize that they have a duty not to deceive, coerce, or exploit their fellow humans (Myskja, 2008). Kant emphasized identifying principles that would be universally true, always, and everywhere. In a study of Chinese enterprises, Chang (2012, p. 427) found that many Chinese people saw the Golden Rule as similar to a principle that exists in Confucian ethics. She quotes a factory worker who says, "People should respect each other. This is a human principle." Thus, she argues that the "general moral imperatives" section of the ACM code (which includes the ideas that one should contribute to society and human well-being, avoid harm to others, promote honesty and trustworthiness, and promote fairness and nondiscrimination) is universal in its scope and its appeal, due to its overlap with Chinese values including loyalty and the Golden Rule (referred to as "the silver rule" in Chinese culture).

A related principle in Kantian ethics states that everyone should treat others as an end in themselves, as people who deserve respect and dignity, rather than merely a means to an end.

Guthrie (2013) uses **deontology** – and the reversibility principle – in describing the ethical responsibilities of technology designers. He argues that "designers should not be attempting to persuade others of something they would not consent to" (p. 57). If you would be uncomfortable with a technology that reads your mail, stores your mail, or monitors your activity online, then you should not be building them or installing them. If you would be uncomfortable with a technology that "prompts" you to do things (not speeding, not leaving the baby in the car on a hot day, not for driving without a seatbelt), then you should extend the same courtesy and respect to other would-be users of this technology.

In the twentieth century, another philosopher (not a moral philosopher, but a political philosopher) took up many of Kant's ideas about the categorical imperative. Like Kant, *John Rawls* believed that people could reason their way to an ethical solution and seek to identify universal rules for behavior that could apply to all people equally. In other words, the value Rawls prized the most was justice, or distributive justice, since he believed that humans had an obligation to seek a just or fair solution to ethical dilemmas. However, he argued that we are not objective in deciding on an ethical solution. When we think about situations and what we should do, we cannot separate our identities as men or women, Americans or other nationals, or rich or poor people. We are tempted to choose the solution in our self-interest. We may believe that we have earned or are entitled to certain advantages. However, Rawls argued that no one should have the right to have more goods or opportunities simply because of any particular character traits or advantages one might have through the circumstances of one's birth.

Thus, Rawls argued that as people reasoned their way to an ethical position or rule, they should engage in a thought experiment to seek what he called *the original position*. Here, they should ask themselves, "If I was blind to my position – i.e., I didn't know my gender, race, social class, nationality – what rule would I then be willing to adopt as universal in this situation?" He referred to this position of not knowing who you were in the scenario as the **veil of ignorance** (Smith, 2016). He suggested that in the reasoning behind the veil of ignorance, people would have to consider the position of the person who was least favored by any proposed hypothetical rule (what he refers to as **the Difference Principle**) and that, in this way, people's regard for those at the bottom of our sociopolitical hierarchy would be strengthened (Douglas, 2015).

Rawls's ideas have influenced the field of philosophy and contemporary politics. Bagnoli argues that Rawls's original position allows people of diverse cultures, religions, and value systems to come together to conclude agreements that would be ethically acceptable to all. Today, much thinking about international governance, the role of foreign aid, and the developed world's duty towards the developing world are influenced by his thinking (Douglas, 2015).

We can contrast Rawls's theory with utilitarianism. Both aim to find the "best" outcome to a dilemma. Still, **utilitarianism** aims to maximize utility in the aggregate (or find the most utility) – regardless of how that utility is distributed. In other words, a solution could be unfair in that some people benefitted significantly more than others, or some benefitted while others lost out. However, if that solution produced the most significant overall utility, it would be considered the best solution. In contrast, Rawls's theory would allow an inequitable solution only if the inequitable solution gave maximum benefit to the least advantaged. It considers not just aggregate utility, but also the distribution of that utility (Schejter, 2007, p. 146).

Critiques of Deontological Ethics

There are many critiques of Kantian ethics. Some analysts suggest that Kant's notions of universal duties in all situations are too idealistic and ultimately unachievable. For example, Kant suggests that humans must engage in peaceful relations and assume that the other is trustworthy.

Others have dismissed Kantian ethics because they see it as inflexible. They suggest that the rule or duty to be truthful should not be absolute. Shouldn't you have the ability to lie to a Nazi official who asked you if you were sheltering Jews during World War II, they ask? Should you tell the truth to someone else whose own intentions are impure and inclined to do ill to you? Others describe this as a misreading of the categorical imperative, noting that Kant did distinguish in his later writings between lying to someone who might harm you (like a criminal) and lying to someone else (Falk, 2005).

Critiques of Rawls

Some of Rawls's critics object to his assumptions in his model. In particular, they disagree with his assumption that the overriding value deciders should wish to pursue is justice. Here some misread Rawls, believing that his emphasis on providing an equitable solution means that he would never countenance a solution in which some people benefitted more. Some people benefitted less from a situation. However, Rawls does acknowledge that a just solution might be one where the wealthy benefitted more, provided that the poor also benefitted and did not lose in any proposed settlement. (Thus, for example, he might find a public building project equitable if it created an infrastructure that allowed everyone to have clean water, even if the project benefitted the wealthy more than it benefitted the poor.)

Writing in *Liberalism and Its Limits*, philosopher Michael Sandel has objected to Rawls's emphasis on acquiring individual goods rather than collective goods. In deciding the equitable solution to a dilemma, Rawls believed that everyone had the right to "equal basic liberties," including the right to vote, to run for office, to have freedom of speech and assembly, and to own private property. He also believed that individuals should have equal opportunities to pursue goods and actions in society (Douglas, 2015). Sandel also argues with Rawls's notion that one can think of an autonomous person who is somehow completely separate from the circumstances of his upbringing, culture, or birth. He goes on to argue that in suggesting that the most well-off should somehow be asked to apply their talents and resources for the benefit of the least well-off rather than for their benefit – in accepting a solution that benefits those at the bottom more than those at the top – Rawls is treating these individuals as a means to an end rather than an end in themselves, which violates the rules that Kant established earlier (Baker, 1985).

Some utilitarian philosophers fault Rawls's use of the veil of ignorance as a deciding mechanism because they feel that choosing the solution that does the least harm to the weakest member is too cautious or risk-averse an approach. Such analysts argue that sometimes one has to take a risk or a gamble in implementing a just solution, and that even if one group suffers in the short term, the good created by a particular policy might be best in the long run. In response, Rawls has argued that it is not rational to gamble with liberties and opportunities (Schroeder, 2007).

Deontological Ethics in Cyberspace

Some analysts make deontological arguments about duties in cyberspace. These arguments assume that there is nothing unique about cyberspace and that individuals have the same obligations and duties to their fellow human that they would have in real space. Spinelli suggests that individuals are obligated to respect intellectual property regimes, since not doing so is a form of theft. He applies the rule of reversibility here, noting that we would not like it if others did not respect our intellectual property. So therefore, we should respect the intellectual property of others. And Fisher and Pappu have suggested that certain practices which occur in cyberspace – such as using bots to generate artificially high numbers of clicks or likes – are forms of deception that should be avoided. They argue that the duty not to lie or practice deception holds in cyberspace just as it does in the real world (Fisher & Pappu, 2006).

In recent years, analysts have also begun asking questions about attribution, trust, and deception in cyberspace, using deontological ethics, including the ideas of John Rawls. Douglas suggests that in thinking about how internet governance might be used to structure a more equitable and just internet for all users – or stakeholders – one might take the original position. How might you feel about net neutrality, surveillance, or rights like anonymity in cyberspace if you did not know if you would be a corporation, an individual user, a person in the developing world, or even someone who did not have access to any internet-connected devices? You might be less motivated to make decisions only in your self-interest or the interests of your company (Douglas, 2015). In her work, Smith has asked who would and would not be helped by requiring engineers to redesign the internet to make it significantly easier to attribute an action (like a cyber-attack) to a specific player. In other words, she asks, "what would a just and equitable internet look like?" (Smith, 2018). Here she references Bishop et al.'s conception of the multiple stakeholders who will be affected by any rule related to attribution: the message sender, the message sender's organization, the sender's government, the ISP the sender uses, the network backbone providers, the government of intermediate nations through which the message passes, the government of the recipient's country, the organization associated with the recipient, and the recipient himself. Thus, they ask what rule might be adopted governing the attribution process, which would not unfairly privilege or penalize any stakeholders (Bishop, 2009). Others have suggested that internet access or access to information should be added to the "equal basic liberties" that Rawls listed (Van den Hoven, 2008).

Box 2.2 Critical Issues: Combatting Disinformation and Fake News

Thus far, we have considered virtue ethics and how it provides a foundation for moral decision-making. But some analysts believe that while virtue ethics should be about embracing or seeking to live according to specific virtues, it should be equally concerned with avoiding or embracing certain vices. In particular, medieval philosopher Thomas Aquinas introduced the notion that there was a corresponding vice for each virtue. For example, we can identify a corresponding vice of selfishness for the virtue of generosity (de Young, 2020).

So what does this have to do with online disinformation? Many ethicists who work in epistemology (or the study and philosophy of knowledge) believe that knowledge is a virtue. Scholars and writers, in particular, should be committing to sharing knowledge with people and making new knowledge.

Therefore, they suggest that a particular vice in epistemology would be engaging in practices that degrade people's knowledge of their world and contribute to people's suspicion about truth and facts. In this view, creating intentionally misleading information is thus an unethical practice.

Defining Disinformation

But what is online disinformation? One standard definition describes disinformation as "false information, as about a country's military strength or plans, disseminated by a government or intelligence agency in a hostile act of tactical political subversion." It is also used to mean "deliberately misleading or biased information; manipulated narrative or facts; or propaganda" (Dictionary.com). Countries have always used disinformation tactics as part of their national defense strategy. If you can mislead your enemies about the size of your military forces or your intentions, you will have the advantage of surprise in a confrontation with them. Thus, disinformation might include producing false or altered photos purporting to show incidents that have not occurred (such as a military attack on a civilian hospital), or even creating false narratives or stories. (For example, during the COVID pandemic, many states attempted to smear their adversaries by suggesting that another country created COVID as a chemical weapon to injure people, including civilians.)

Why Is Disinformation Wrong?

Disinformation thus is a type of lying or deception. While military ethics sometimes allows for the use of deception as a wartime tactic, a virtue ethics perspective that emphasized honesty as a virtue would fault disinformation tactics as a type of dishonesty or dishonest behavior.

However, one can make additional philosophical arguments about disinformation. The philosopher Luciano Floridi describes the "infosphere" as an environment in his work. Just as it's essential to preserve our physical environment from harms like pollution, he suggests that we should feel morally obligated not to harm our infosphere or information environment by perpetrating acts like spamming or disinformation. He suggests that the infosphere should serve as a trusted source of information. Just as we are fortunate enough in the developed world to trust the quality of the water from the tap in our kitchen, he suggests we should be able to trust the information circulating in the infosphere (Floridi, 2014). But acts like disinformation cause us not to trust the information we encounter online. Instead, we are suspicious and distrustful. We may distrust the information in a particular story and even eventually distrust any online information as a source.

Here the analysts Baird and Calvard employ the idea of epistemic vice to describe the actions of those who produce disinformation and those who allow it on their platforms. They point to platforms like Facebook – in which creators of false content can receive and solicit payment for creating false content that nonetheless performs well (i.e., receiving lots of likes and shares) and paying Facebook to promote fake stories. In their view, Facebook and other platforms do not seek the "epistemic well-being of stakeholders" and platform users. (That is, they are not primarily interested in having an audience that is well informed and which has access to truthful, trustworthy content.) Here, Baird and Calvard compare platforms like Facebook to tobacco companies that knowingly misled smokers in the 1970s and 1980s about the dangers posed by their products. In both instances, they argue that the primary ethical commitment was to disseminating falsehoods rather than truth, even if falsehoods ultimately hurt people and organizations (Baird, 2020).

In this view, individuals like Mark Zuckerberg and corporations like Twitter have an ethical responsibility to furnish accurate information to their users and investigate situations in which inaccurate information is being disseminated. Pursuing the virtue of truth and avoiding the vice of dishonesty and false information is seen as a greater good – outweighing any concerns that users might raise about censorship or control on a platform.

Finally, some technology scholars have begun to think about what it means to "design for democracy." How can the design of an interface, for example, affect users' trust in the material presented? How might users be led to investigate multiple views or widen their search for information about a topic? Scholars note that ideally, designing for democracy would be a challenge that leaders at social media companies like Twitter and Facebook would also embrace. Designing for democracy rests on the principle that disinformation is, to some degree, a social problem that can be combatted through a better understanding of how users pursue and organize information and by helping them in their search for trustworthy, objective information (Tonkinwise, n.d.).

Bibliography

Baird, C. A. (2020). Epistemic vices in organizations: Knowledge, truth and unethical conduct. *Journal of Business Ethics*, *160*(1), 263–276.

de Young, R. (2020). *Glittering vices: A new look at the seven deadly sins*. Grand Rapids, MI: Brazos Press.

Dictionary.com. (n.d.). Disinformation. Retrieved October 3, 2022, from https://www.dictionary.com/browse/disinformation.

Floridi, L. (2014). *The fourth revolution: How the infosphere is reshaping human reality*. Oxford: Oxford University Press.

Tonkinwise, C. (n.d.). Democracy must be defended – by undemocratic designs. *Academia.edu*. Retrieved June 1, 2022, from www.academia.edu/35779909/Democracy_must_be_Defended_by_Undemocratic_Designs.

Deontological ethics has also affected thinking in robot ethics. As Yu points out, duty-based ethics requires that the decider can reflect upon their actions, ask "what would happen if everyone was able to behave in the way in which I am behaving in every circumstance," and ask, "Am I treating the others in this situation as merely a means to an end?" Thus, he asks whether machines could ever be taught to think morally through deontological ethics since they cannot reflect upon their actions in the same way humans can (Yu, 2012). For this reason, analysts like Arkin and Wallach have suggested that if one believes that machines can be taught to think ethically, it is more likely that they can be taught to calculate the utility of a particular decision to maximize this outcome. Machines could eventually learn to think morally using utilitarian ethics, but they are unlikely to become virtuous or comprehend their moral duties using deontological ethics (Arkin, 2009; Wallach, 2010).

What Are Utilitarian Ethics?

The third lens is utilitarianism. Utilitarianism is sometimes referred to as a teleological or consequentialist theory since it is concerned with the endpoint or the decision's consequences rather than the decision-maker's attitude or intent. Utilitarianism is newer than virtue ethics. We can trace it back to the ideas of *Jeremy Bentham* (1748–1832), an eighteenth-century British social reformer. The practical framework arose from the *Enlightenment*, a period in European history in which society was captivated by the notion that reason or logic could guide social behavior in the world. The reason was often explicitly described as the opposite of a religious sensibility. Authors thus stated that while religion often told people what to do based on scriptures, religious teachings, and centuries of moral practices in society, reason allowed individuals to choose for themselves what the best course of action might be based on more scientific principles. In particular, Bentham argued for a decision-making calculus based on hedonism or the pursuit of pleasure. He argued that one should always seek to achieve pleasure and avoid pain. These rules could then be used to govern what we ought to do. He argued that we could arrive at an ethical decision based on reason and law rather than looking to religion or any form of higher order. For this reason, we often use the phrase "utilitarian calculus" to refer to a process of decision-making in which individuals weigh the possible costs and benefits associated with a particular choice.

In his essay, *John Stuart Mill* (1806–1873) built upon Bentham's ideas, called "Utilitarianism." Here he claimed that he wasn't inventing a theory that people would use in decision-making. Instead, he offered a historical look that argued that people unconsciously act like utilitarians, totaling up the possible costs and benefits of undertaking one course of action rather than another. He traces this line of thinking back to the Ancient Greeks. He also argues that the "rules" which Kant identifies in his deontological ethics, such as "act in the same way that you would want others to act towards you," are in reality based on utility theorizing, since it would make little sense for us to mistreat our opponents if we knew that someday we might be in a situation where they would do the same towards us.

Comparing Virtue Ethics and Utilitarianism

Thus, we can explicitly contrast utilitarianism with virtue ethics, since virtue ethics assumes that the outcome alone is not the most critical determinant in deciding the moral choice. Virtue ethics states that one's orientation towards a subject is essential, and that it is as essential to wish to do good as it is to have a good outcome. Virtue ethics also suggests that an individual's ethical obligations – to act reasonably and with good intent – are essentially unchanging since one's character is regarded as unchanging.

Utilitarianism, in contrast, also allows for the possibility of *situational ethics*. Situational ethics suggests that you might need to violate society's or your own moral code to provide the most moral outcome in some circumstances. In practical ethics, the "best" decision is the highest payoff or creates the most utility or happiness. Utilitarian ethics also acknowledges that there will be tradeoffs. To create utility for the many, it may be necessary to create disutility or to sacrifice the needs of one or more individuals. Here we can think back to the Trolley Problem. The utilitarian would agree that sometimes pushing the fat man off the bridge is necessary to stop the trolley from injuring the children playing on the tracks. The utility derived from saving several children is, in this case, more significant than the loss incurred through the death of the fat man.

Pros and Cons of Utilitarian Ethics

Utilitarian ethics have much to recommend. Compared to other ethical lenses, the practical model is *parsimonious* – i.e., it explains many things using a relatively simple mechanism. It is also seen as morally neutral since it can be applied objectively, without regard to one's underlying beliefs or culture. It could be seen as universally valid across cultures and periods. Here Mill shows that utilitarianism usually lines up with conventional morality.

Indeed, Bonnefon et al. believe utilitarian ethics could be programmed into self-driving autonomous vehicles (AV) driving programs. They argue that it is possible to create "moral algorithms that align with human moral attitudes" (Bonnefon et al., 2016). For example, a programmer could instruct a vehicle to minimize the death toll if a vehicle crash is likely. While an AV cannot monitor its intent (as a human is expected to in practicing virtue ethics), it could calculate likely outcomes using a decision tree and arrive at a "moral choice" using a utilitarian calculus. Even a self-driving car has the potential to calculate which is the lesser of two evils.

However, there are critics of the theory. First, ethicists have asked if maximizing utility is the same as doing what is good or right. Here, we can imagine actions that a practical approach might prescribe that seem morally or ethically troublesome. For example, imagine a situation where a single man with no dependents is an exact match for an organ donation to a man who has a large family and several people depending on him. One could argue that the most significant amount of happiness would be achieved if the family with many children was allowed to continue to have a healthy father; therefore, the moral thing to do would be to force the single man to donate his organs, even if doing so would kill him. One could also argue that it is moral to force everyone who earns a certain amount of money to donate to charity. This would maximize happiness for the most significant number of people who would likely receive benefits. In each instance, the individual's rights are subordinated to the group's rights, and the proposed solution can be seen as authoritarian (West, 2004, p. 23).

Military ethicist Edward Barrett (2013, p. 4) takes this approach, arguing that practical thinking is inappropriate for considering the ethics of cyberwarfare. He argues that individual rights to liberty and dignity "cannot be overridden by a consequence-derived utilitarian calculation" and thus recommends applying a just war perspective to thinking about ethics in cyberspace.

Utilitarian Ethics in Cyberspace

In applying utilitarianism in computer ethics, we again encounter the **uniqueness debate**. Should we have different norms and morals for dealing with one another in cyberspace from what we have for dealing with one another in real space? And finally, how do we define "the good" in cyberspace? Is it the same as "the good" in meat space, or is it somehow different?

Among computer ethicists, Moor (1998, 1999), one of the original framers of the ACM Code of Ethics, strongly promotes a "just consequentialism." He argues that only a universal set of computer ethics is proper and cautions against framing a culturally specific or relative set of ethics. Instead, he feels we can identify core ideas in computer ethics to frame a universal ethical theory. Moor introduces the acronym ASK FOR. He states, "no matter what goals humans seek, they need ability, security, knowledge, freedom, opportunity, and resources to accomplish their projects. There are the kinds of goods that permit us to do whatever we want to" (Moor, 1999, p. 66). Thus, he suggests that decision-makers embrace policies that allow individuals to have these goods and avoid the "bad" described earlier (death, pain, disability, etc.). Here he also cautions against choosing a policy that provides maximum good in the short run but may create unpleasant consequences in the long run. He asks the reader to imagine a situation where a marketing corporation can buy a database with vast amounts of detailed personal information about everyone in a country. The good is apparent, he argues. The company can sell more products. But in the long run, people will have less autonomy, and freedom, as more details about their lives are known (Moor, 1999, p. 67).

In their work, Tuffley and Antonio grapple with how we define "the good" regarding information technology. They conclude that:

> In the broadest sense, technology is ethical when it is life-affirming, when it helps people grow towards their full potential when it allows them to accomplish what they might otherwise not be able to.
>
> (Tuffley & Antonio, 2016, p. 20)

Here the claim is that technology's goal is to produce human flourishing, which is the standard by which it ought to be measured. Using this yardstick, a computer engineer can ask, "What is the likelihood that more or greater human flourishing would be produced by choosing this action over another?" Here, intent or psychological attitude does not matter; the outcome is primary, as the utilitarian ethics approach prescribes.

Comparing and Contrasting the Models

In this chapter, we encountered three models for thinking about computer ethics. The agent-centered virtue ethics model assumes that individuals make decisions and that the most ethical solution is the one that helps an individual to develop their character, leading to human flourishing. This model is the oldest and is rooted in history and tradition, including religious traditions. It is attractive to individuals today because of its emphasis on each individual's inherent rights and dignity and its requirement that decisions be made and applied consistently. That is, "the right thing to do" does not vary according to the place or position in which one finds oneself.

The utilitarian model, in contrast, advocates a type of situational ethics where "the right thing to do" is highly affected by the environment in which decisions are made. While murdering another human being is seldom "the right thing to do," a utilitarian might argue that if the person is a future tyrant like Hitler or Mussolini, it might be ethically appropriate to destroy the individual to achieve an outcome that is better for all. This model purports to be highly rational and universal – in the sense that one can perform calculations regarding utility in any nation or any culture simply by adding up the costs and benefits of particular courses of action. In this model, as noted, the decision-maker's intent is not essential. Instead, what counts is the outcome.

Finally, the deontological model considers who might be affected by an ethical decision. Which choice allows the most humane treatment of the participants, not treating them as a means to an end but as ends themselves? This model helps us think about technology's effects on people and how people are affected by technological decision-making. This model acknowledges that humans have moral duties and that their duties are to one another.

As we have seen, the models require the decision-maker to ask a different series of questions before making a decision. Table 2.1 lays out the decision-making calculus for each approach to "doing ethics":

In the following chapters, we apply the frameworks to think through problems in cybersecurity ethics. We use the frameworks to consider privacy, surveillance, military ethics, and intellectual property. As Table 2.2 indicates, each model or lens highlights certain facets of the issue being considered while downplaying others. As we shall see in the following chapters, each has strengths and weaknesses.

Table 2.1 Applying the Models

Model	*Questions to Ask in Ethical Decision-Making*
Virtue Ethics Approach	• Which position best expresses my values and character? • If I choose this, can I live with myself? • Will it contribute to my own human flourishing/character development?
Utilitarian Approach	• Which position generates the greatest positive utility and produces the fewest negative consequences? • What costs are associated with each outcome? • What benefits are associated with each outcome?
Deontological Approach	• Who will be affected by this decision? • Am I treating others as a means or an end in themselves? • If my actions became a rule and I was subject to that rule, would I accept it and view it as ethical?

Table 2.2 Comparison of Ethical Frameworks

	Pros	*Cons*
Virtue Ethics	• Ethical positions are consistent across issues • Allows for solutions to new and novel ethical dilemmas • Emphasizes character of decision-maker	• Compassion problem • Problem of Evil • "Traditional" – and therefore perhaps outdated in new environments
Utilitarian Ethics	• Calculations are clean and value-free • Can be taught • Used in AI, with robots • Universally valid rather than idiosyncratic or changing with culture	• There are no absolute moral imperatives • Leads to an instrumental view of human beings – as a means to an end rather than valuable in their own right
Deontological Ethics	• Focus on those affected by the decision • Reciprocal – forces agent to see herself as both the decider and subject of decision	• Overemphasizes duty to individuals over duty to produce best possible outcome • Can be inflexible in insistence on universal rules • Can be stodgy and risk-averse (Rawls)

CHAPTER SUMMARY

- Over time, different ways of evaluating the ethical aspects of actions have emerged, based on developments in religion and philosophy.
- The three frames considered here (virtue ethics, utilitarianism, and deontological ethics) differ in terms of what constitutes ethical behavior. Is it acting in line with one's values? Achieving a particular ethical outcome or acting in line with moral standards?
- Virtue ethicists believe that there is some objective list of virtues that, when cultivated, maximize a person's chance of living a good life.
- Utilitarianism assumes that one can measure the utility of particular choices and decide rationally which action will yield the most utility. In seeking a particular end, other values may need to be compromised, and it assumes that what is "best" in one situation might not be best in another.
- Deontological ethicists believe that humans can use their reasoning abilities to derive an ethical position by asking a series of questions, including "What would be the outcome if everyone acted this way?" And, "Would I approve of this behavior having the status of a universal law?"
- Each framework allows us to make a particular type of ethical argument. The frameworks might not agree on the best choice in a particular situation.

DISCUSSION QUESTIONS

1 **Risk, Safety, and Information Sharing**

Students may wish to visit the website of the Electronic Frontier Foundation, a nonprofit organization concerned with user privacy. Read this article about the Ring doorbell app and the problems which EFF has identified: www.eff.org/deeplinks/2019/08/amazons-ring-perfect-storm-privacy-threats/ (accessed June 21, 2022).

What are the ethical issues you see referenced in this article? Describe the ethical principles presented and how they may conflict with one another.

2 **Competing Values and Context**

The Eurobarometer Survey is a large-scale public opinion questionnaire conducted on an ongoing basis throughout the European Union. A recent version of the Eurobarometer asked citizens in several European countries about their values and which values were most important to them. Take a look at the summary from this survey, found here: https://europa.eu/eurobarometer/surveys/detail/2230

Now imagine that you are being asked to design an app used in all schools in the European Union to track students' grades, including whether they are completing work on time. The proposal for the app states that it will help parents better understand their student's learning needs and help school districts (nationally and locally) better support struggling students.

Consider the data you would and would not want to collect in designing this app. (For example, would you need to know students' immigration status, whether they speak a different language at home, or their parent's education level?)

- How can you use the information about citizen values from the Eurobarometer survey to design the app?
- What information might European users want to collect, store, and share, and what information might they wish not to collect, store, and share?

- Who might users want to share this information with, and who might not want to share it?
- Should information be shared automatically, or should users consent to share information?
- What might be the downsides of designing such an app, and what might be the advantages?

RECOMMENDED RESOURCES

Douglas, D. M. (2015). Towards a just and fair internet: Applying Rawls's principles of justice to internet regulation. *Ethics of Information Technology*, *17*, 57–64.

Edmonds, D. (2015). *Would you kill the fat man? The Trolley Problem and what your answer tells us about right and wrong*. Princeton, NJ: Princeton University Press.

Jaipuria, T. (2015, May 23). Self-driving cars and the Trolley Problem. *Medium.com*. Retrieved June 1, 2022, from https://medium.com/@tanayj/self-driving-cars-and-the-trolley-problem-5363b86cb82d.

References

Afthanassoulis, N. no date. "Virtue Ethics." In Internet Encyclopedia of Philosophy. Availalbe at www.iep.utm.edu/virtue/. Accessed June 14, 2016.

Afthanassoulis, N. (2012). *Virtue ethics*. London: Bloomsbury.

Ali, A. (2014). *Business ethics in Islam*. London: Edward Elgar.

Arkin, R. (2009). *Governing lethal behavior in autonomous robots*. New York: Chapman & Hall.

Baker, C. E. (1985). Sandel on rawls. *University of Pennsylvania Law Review*, *133*(145), 895–928.

Barrett, E. (2013). Warfare in a new domain: The ethics of military cyber operations. In T. Demy & G. A. Lucas (Eds.), *Military ethics and emerging technologies* (pp. 4–17). London: Routledge.

Barriga, A. Q. (2001). Moral cognition: Explaining the gender difference in antisocial behavior. *Merrill-Palmer Quarterly*, *47*(4), 532–562.

Bishop, M. G. (2009). The sisterhood of the traveling packets. In *NSPW Proceedings of the 2009 workshop on new security paradigms* (pp. 59–70). Oxford: New Security Paradigms Workshop.

Bonnefon, J.-F., Shariff, A., & Rahwan, I. (2016). The social dilemma of autonomous vehicles. *Science*, *352*(6293), 1573–1576.

Borowitz, E., & Schwartz, F. W. (1999). *The Jewish moral virtues*. Philadelphia, PA: The Jewish Publication Society.

Brey, P. (2007). Ethical aspects of information security and privacy. In N. A. Petkovic (Eds.), *Security, privacy and trust in modern data management* (pp. 21–36). Heidelberg; Berlin: Springer.

Chang, C. L.-H. (2012). How to build an appropriate information ethics code for enterprises in Chinese cultural society. *Computers in Human Behavior*, *28*(2), 420–433.

Douglas, D. M. (2015). Towards a just and fair internet: Applying Rawls' principles of justice to internet regulation. *Ethics and Information Technology*, *17*(1), 57–64.

Falk, C. (2005). *The ethics of cryptography*. West Lafayette, IN: Purdue University.

Fisher, J., & Pappu, R. (2006). Cyber-rigging click-through rates: Exploring the ethical dimensions. *International Journal of Internet Marketing and Advertising*, *3*(1), 48–59.

Gottlieb, P. (2009). *The virtue of Aristotle's ethics*. Cambridge: Cambridge University Press.

Guthrie, C. F. (2013). Smart technology and the moral life. *Ethics and Behavior*, *23*(4), 324–337.

Jaipuria, T. (2015, May 25). Self-driving cars and the Trolley Problem. *Medium.com*. Retrieved June 2, 2022, from https://medium.com/@tanayj/self-driving-cars-and-the-trolley-problem-5363b86cb82d.

Koller, P. (2014). On the nature of norms. *Ratio Juris*, *27*(2), 155–175.

Kouchaki, M., & Smith, I. H. (2014). The morning morality effect: The influence of time of day on unethical behavior. *Psychological Science*, *25*(1), 95–102.

Lessig, L. (2000). *Code: And other laws of cyberspace*. New York: Basic Books.

Moor, J. H. (1998). Reason, relativity and responsibility in computer ethics. *Computers and Society*, *28*(1), 1–16.

Moor, J. H. (1999). Just consequentialism and computing. *Ethics and Information Technology*, *1*(1), 65–69.

Myskja, B. K. (2008). The categorical imperative and the ethics of trust. *Ethics and Information Technology*, *10*(1), 213–220.

Olivier, B. (2013, July 28). Is there a need for cyber ethics? *Thought Leader*. Retrieved June 2, 2022, from http://thoughtleader.co.za/bertolivier/2013/07/28/is-there-a-need-for-cyber-ethics/.

Schejter, A., & Yemini, M. (2007). Justice, and only justice, you shall pursue: Network neutrality, the first amendment and John Rawls's theory of justice. *Michigan Telecommunications and Law Review*, *14*(1), 137–174.

Schroeder, D. (2007). *Rawls and risk aversion*. Unpublished Manuscript. Retrieved from www1.cmc.edu/pages/faculty/ASchroeder/docs/Rawlsmaximin.pdf.

Slingerland, E. (2011). The situationist critique and early confucian virtue ethics. *Ethics*, *121*(1), 390–419.

Smith, J. (2016, November 14). Attribution from behind the veil of ignorance. *The National Interest*. Retrieved from https://nationalinterest.org/feature/the-great-cybersecurity-attribution-problem-18385.

Smith, M. (2018, March 27). *Online presentation: Why is internet access still considered a luxury in America?* Cambridge, MA: Harvard University Berkman Klein Center for Internet and Society. Retrieved July 25, 2022, from https://cyber.harvard.edu/events/2018/luncheon/03/Smith.

Tuffley, D., & Antonio, A. (2016, January–March). Ethics in the information age. *Australian Quarterly*, 19–24.

Vallor, S. (2013). The future of military virtue. In K. S. Podins (Ed.), *5th international conference on cyber conflict*. Tallinn, Estonia: NATO CCD COE Publications.

Van den Hoven, J. (2008). Information technology, privacy and the protection of personal data. In J. A. Van den Hoven (Ed.), *Information technology and moral philosophy*. New York: Cambridge University Press.

Walker, R., & Ivanhoe, P. (2007). *Working virtue*. Oxford: Clarendon Press.

Wallach, W., & Allen, C. (2010). *Moral machines: Teaching robots right from wrong*. Oxford: Oxford University Press.

Walsh, A. (2011). A moderate defense of the use of thought experiments in applied ethics. *Ethical Theory and Moral Practice*, *14*(4), 467–481.

West, H. (2004). *An introduction to Mill's utilitarian ethics*. Cambridge: Cambridge University Press.

Yu, S. (2012). *Machine morality: Computing right and wrong*. New Haven, CT: Yale Scientific.

3 The Ethical Hacker

LEARNING OBJECTIVES

At the end of this chapter, students will be able to:

1 Define hacking and describe how computer hacking has changed and evolved since its inception
2 List at least five reasons why hackers engage in hacking activities, and describe the types of hackers
3 Compare and contrast the conditions and approaches of white hat, grey hat, and black hat hackers
4 Describe the licensing mechanisms currently used for certifying an ethical hacker
5 Provide an evaluation of the ethics of hacking and penetration testing using virtue ethics, utilitarian ethics, and a deontological framework

In considering the ethics of hacking, we need to consider several factors – including who the hack targets, the intents of the hacker, and the conditions under which the hack takes place.

As we consider the ethical problem of hacking in this chapter, we can consider five real-life situations involving a variety of actors (from individuals to nation-states) and in pursuit of a variety of ends – from the theft of money and resources to attempts to compromise critical infrastructure, to hacking for espionage.

- In August 2021, Italian hackers shut down the system for booking COVID-19 vaccination shots in Italy's Lazio Region. Italy had begun requiring proof of vaccination for participation in many social activities, and large-scale protests had erupted (Reuters, 2021).
- In 2021, a group of hackers began offering "banning as a service." For $60, users could hire them to get anyone they chose banned from the Instagram platform. Hacks involved creating fake posts threatening suicide and self-harm through impersonating the users (Cox, 2021).
- In March 2021, the European Banking Authority's Microsoft Exchange servers were compromised due to a hacker attack. This organization oversees the orderly functioning of banks within the European Union.
- In March 2020, the technology vendor Solar Winds was the subject of a nation-state-sponsored hacker attack by Russian intelligence. The attack affected government and private actors, including the energy and communications sectors throughout Europe, North America, Asia, and the Middle East. The aim of this attack was likely

DOI:10.4324/9781003248828-4

intelligence collection since it compromised the US Department of Homeland Security and Departments of Justice, Commerce, Treasury, Energy, and possibly the Pentagon. The attack went undetected for eight months.

- In April 2021, the DarkSide gang attacked Colonial Pipeline's billing system and internal business network, launching a ransomware attack. This hack disrupted gas supplies on America's East Coast, leading people to engage in hoarding and other disruptive actions. Colonial Pipeline paid a $4.4 million ransom, eventually recovered by the US Federal Bureau of Investigation. This event represented an attack on critical infrastructure.

What do these stories have in common? Someone with specialized knowledge gained access to a system in each instance, outsmarting and outwitting any defenses created against entry. But the stories are also very different. Hackers may be seeking to change a policy, capture resources, or harm individuals or a state entity. As the examples show, hacking can have a variety of repercussions – emotional and social, financial, legal, and political.

This chapter considers what hacking is and is not, how the so-called hacker ethic has changed and developed over time as computer security has become more professionalized, and the differences between white hat, grey hat, and black hat hacking.

What Is a Hacker?

A *hack* refers to an unconventional way of doing something. When the term originated in the 1980s, it was often applied to individuals like Steve Jobs and Bill Gates. They were seen as engaging in heroic and creative endeavors that produced new technological resources which enriched society.

However, today hacking has both negative and positive meanings. Today, the term is often used to denote someone who wishes to gain *unauthorized access* to a system (for example, if an individual working in one section of a corporation is looking at files belonging to another section of the corporation for which he had not been given permission) or it might mean gaining *illegal access* to a system (through utilizing stolen passwords, impersonating another user, or simply using an algorithm to guess a password).

While hacking may have initially begun in the early days of the computer revolution as merely a game or joke, today, hacking is often viewed as a type of *cybercrime* or used as a tool to commit a traditional crime. Hobbyist hackers may offer their skills to the highest bidder, and syndicates run by cybercriminals may often purchase those skills. Today's hackers may engage in acts considered to be vandalism, destruction of property, and theft.

Today, most analysts agree that certain activities that fall broadly under hacking can also be described as cybercrime variants. Here, Tavani (2004, p. 121) includes:

Cybertrespass – the use of information technology to gain unauthorized access to computer systems or password-protected sites

Cyber vandalism – the use of information technology to unleash programs that disrupt the operations of computer networks or corrupt data

Computer fraud – the use of deception for personal gain in online business transactions by assuming a false online identity or by altering or misrepresenting data

Cyberterrorism is the execution of politically motivated hacking operations intended to cause grave harm, resulting in either loss of life or severe economic loss.

Cybercrimes may have a variety of targets – from private actors to corporate actors to international organizations and nation-states. Hackers may also target different components within a computer system.

Databases can be attacked via an SQL injection attack. Here attackers submit a malicious SQL (or structured query language) statement to the target. Since SQL can be used to search data, erase or dump data, or alter data, a successful query can alter a target database and permit administrator access to the database and possibly other parts of a host operating system.

Meanwhile, a supply chain attack rests on the fact that today many organizations do not supply and service all of the components in their supply chain. A company may outsource some of the tasks it needs to perform, such as storing its data, performing payroll tasks, and managing its e-mail and communications. An attack on some component of that supply chain (such as a data storage cloud) can have widespread effects on many companies that depend on that component. An attack on this component may offer the attacker access to multiple targets. (The Solar Winds attack noted earlier was a supply chain attack.)

Cybercrimes may include attacks on hardware or software (or the supply chain) to damage a system or data. Still, cybercrime might also include "common crimes," which rely on technology to achieve their effect. (That is, one can either steal a physical credit card or steal someone's credit card data in cyberspace. Both represent theft, but the second instance is a cybercrime since a computer was used to carry out the theft.) Hacking and the use of technology can thus be used to carry out additional crimes like cyberstalking and cyberbullying. Here, information gathered through the internet might enable an attacker to know where to find a victim or identify the victim's vulnerabilities.

Thus, hacking is often unlawful. Specific hacks violate a variety of federal and state laws. In the United States, the *Computer Fraud and Abuse Act (CFAA)* spells out the federal punishments related to hacking, including imprisonment of up to 20 years. Other US federal laws include the Wiretap Act, the Unlawful Access to Stored Communications Law, the Identity Theft and Aggravated Identity Theft Laws, the Access Device Fraud Law, the CAN-SPAM Act, and the Communication Interference Act. The Electronic Communications Privacy Act (ECPA) establishes ransomware attacks as a federal offense. We will examine these legal issues in more detail in subsequent chapters on privacy and surveillance.

Box 3.1 Tech Talk: Is Web Scraping Considered Hacking?

Web scraping is a process by which companies or individuals use automated programs (or bots) to impersonate humans who might browse the web. These automated programs can go through a website, clicking on information and links. They can then store the information collected in databases, collecting large sets of information about various topics. If you have ever typed a query into Google such as "real estate listings near me," you may have been directed to a site displaying information obtained through web scraping.

Web Scraping Is an Ethical Breach

But is web scraping illegal, and is it a form of hacking? Using a web scraping program to impersonate or imitate a human user can be considered deception and an ethical breach. Users must often sign the Terms of Service (TOS) before using a site.

Terms of Service often contain language in which the user acknowledges that she is accessing the site as an individual to use the information personally – rather than as a corporation to collect large-scale information for corporate use.

Facebook's TOS expressly forbids users from engaging in Automated Data Collection, which they define as "the collection of data from Facebook through automated means such as harvesting bots, robots, spiders or scrapers." Without Facebook's express permission, people cannot engage in Automated Data Scraping. Similarly, the job-hunting site LinkedIn argues that a user who uses a bot to, for example, collect a list of programmers who have a specific certification for employment recruiting is using the site commercially rather than for private use. Thus, in June 2020, LinkedIn sued the workforce analytics company hiQ Labs, Inc. for violating its terms of use by engaging in web scraping (Vittorio, 2021).

Other ethical critiques of web scraping focus on anonymity breaches that might occur if web scraping leads to collecting personally identifiable information, mainly if multiple datasets obtained through web crawling are aggregated. European Union officials voiced concerns after data scraped from Facebook and LinkedIn was leaked and made public. Both platforms and companies that scraped the data were accused of violating the European Union's General Data Protection Regulation (GDPR). This regulation requires that users consent when their data is collected, and web scraping bypasses the consent mechanism.

Web scraping might also be viewed as theft of a company or platform's resources. A web scraper clicks on every link on a page, whereas the average user does not. Therefore, it can slow down or crash an entity's server and prevent legitimate users from doing their work. Developers spend money and time developing bot detection systems such as CAPTCHA codes, and users spend their time proving that they are not bots (Li et al., 2021).

And because bots that engage in web scraping work so quickly, they tend to "win" in human interactions. A web scraping bot that procures information about the availability of real estate, concert or train tickets, or goods for sale at an auction site like eBay can contribute to the inequitable distribution of resources. In 2020, Indian authorities launched a legal case against the Tatkal Ticket Software program Jaguar, arguing that it was using illegal technology to evade CAPTCHA mechanisms to unfairly purchase more than its share of tickets on India's railway system.

Regulating Web Scraping

In the introduction to this textbook, we described how technological developments sometimes outpace our ability to regulate them legally. At present, web scraping exists in a legal **grey area** – it is neither wholly legal nor illegal, and there are no explicit norms or social expectations regarding what is and isn't allowed. The targets of web scraping are often not government entities but commercial sites such as LinkedIn, eBay, Twitter, and Google, to complicate the matter. Therefore, much of the regulation and enforcement rests on the actions not of governments, but of private actors.

The Role of Private Corporations in Regulating Web Scraping

Today, sites like Facebook and Google offer researchers access to their collected information, most commonly through providing an **Application Programming Interface (API)**. An API is a software program that allows applications to talk to each other. The API formats and delivers the request to a provider (like Facebook) and then the data back to the client. Until recently, Facebook offered an API to all users. Today, however, prospective clients must receive permission to access the API. A committee considers their request at Facebook – essentially playing a role similar to that of a university's Institutional Review Board, ensuring that data protection and anonymization protocols are in place and carried out.

In this way, web scraping represents a sort of "back door" to retrieving data – rather than going through the front door provided by the API. It has been compared to "trespass," and some legal scholars suggest that laws about "trespass to chattel" could be invoked to punish web scrapers. Trespass to chattel refers to a situation where a person interferes with someone else's property and, in the process, harms that property. In addition, some platforms have sought to ban individuals or groups which have engaged in web scraping from accessing their platforms (Fisler, Beard, & Keegan, 2020).

Bibliography

Fisler, C., Beard, N., & Keegan, B. C. (2020). *No robots, spiders or scrapers: Legal and ethical regulation of data collection methods in social media terms of service*. Palo Alto, CA: Association for the Advancement of Artificial Intelligence.

Li, X., Azad, B., Rahmati, A., & Nikiforakis, N. (2021). Good bot, bad bot: Characterizing automated browsing activity. *Securitee.org*. Retrieved June 1, 2022, from https://amir.rahmati.com/publication/aristaeus-ieeesp21/.

Vittorio, A. (2021, May 4). Scraping episodes highlight debate over anti-hacking law's scope. *Bloomberg Law*. Retrieved June 1, 2022, from https://news.bloomberglaw.com/privacy-and-data-security/scraping-episodes-highlight-debate-over-anti-hacking-laws-scope?_pxhc=1650426552103.

In addition, hacking is mentioned in the *USA Patriot Act*, legislation on US national security that was passed in the aftermath of the 9/11 terrorist attacks. The Patriot Act notes that breaking into federal computers may fall under the definition of terrorism, and cyberhackers can be prosecuted as terrorists (Young et al., 2007, p. 281). On a state level, hacking behaviors may also violate laws regarding phishing and the use of spyware (hackerlaw.org). Similarly, in Great Britain, the Terrorism Act of 2000 listed hacking or threatening to engage in computer hacking as a potential terrorist act. If the hacker was doing so in support of a political, religious, or ideological cause, or in order to influence the government or intimidate the public or some segment of the public, it is considered terrorism (Inbrief.co.uk). According to the Criminal Damage Act 1971 and the

Computer Misuse Act, hackers in the UK can also be prosecuted (www.inbrief.co.uk/offences/hacking-of-computers).

Is Hacking Always Wrong?

However, it is too simplistic to say that all hacking is consistently ethically wrong or always ethically justified. Similarly, not all hacking is illegal. Instead, as we will see in this chapter, we need to consider several factors – from who the hack targets, to the intents of the hacker, to the conditions under which the hack takes place.

Bratus et al. (2010) define hacking as merely a package of skills that computer experts can develop, arguing that "Hacking is the skill to question trust and control assumptions expressed in software and hardware, as well as in processes that involve human(s)-in-the-loop (a.k.a. 'Social Engineering')." Thus, hackers acquire and use skills like the ability to encrypt and decrypt data, create and transmit viruses, and identify and diagnose security vulnerabilities within a computer system. They argue that just like doctors, locksmiths, or martial artists could use their skills to aid humans or harm them, hackers are not constrained to act ethically or unethically. It is moral reasoning, they argue, in the last analysis that will determine how hackers use the skills that they have developed.

At the same time, many hackers argue that users have robust security and are therefore responsible when their systems are hacked, not the hackers. Indeed, some hackers argue that they are performing a public service in letting companies know that their systems are vulnerable.

However, Xu et al. caution that while hackers might start innocently, they may be drawn into engaging in less ethical pursuits. These authors stress that the best predictor of whether someone will continue to engage in mischief or move on to more harmful pursuits is their ability to engage in moral reasoning (Xu et al., 2013). Students who understand that their actions have ethical consequences can, they argue, set limits regarding what behaviors cross an ethical line.

Why Do People Hack?

Considering the four cases encountered at the beginning of this chapter, we see that not all hackers have the same intentions and that the consequences of a hack can vary greatly. And ethics and law do not always line up neatly. Some actions are both illegal and unethical (i.e., the Solar Winds attack), and some are arguably illegal yet ethical (i.e., "breaking into" a cell phone that does not belong to you if you are a law enforcement officer investigating a crime), and some might be legal but still unethical (i.e., perhaps "cheating" on a video game.).

In considering the types of hacking behavior, we can place acts on a spectrum ranging from least to most harmful. As we move from least to most severe, we can see that the type of person liable to engage in the activity changes, as do the types of attacks, the level of organization of the groups, and the overall destructiveness of the acts themselves. Table 3.1 illustrates the various types of hacks possible today, in increasing order of severity.

As Table 3.1 indicates, hackers may have more than one motive, and a hacking attempt may be a "one-off" event or related to a larger strategy. In particular, it is becoming more difficult to distinguish between hacks like political **doxing**, which might include releasing embarrassing information about a political official, and acts of war. As we saw in the

Table 3.1 Types of Hacking

Type of Attack	*Examples*	*Motivation*	*Target*	*Type of Attacker*
Nuisance	• Cheating on a video game • Playing pranks	• Demonstrate skill • Enhance reputation • Learning	Individuals: general or selected	Least well organized – may be students or youth
Activist	• Attempting to change the outcome of political events – leaking information, tampering with elections, political doxing	Ideological	Public figures Corporations	May be a small or large organization; may be a spontaneous or permanent organization
Criminal	• Identity theft • Theft of assets • Installation of spyware, malware, and ransomware	• Money • Profit	Individuals Rival corporations	May be an individual or a criminal syndicate; ties to organized crime
Act of War	• Attacks on critical infrastructure • Changing coordinates for military targeting • Acts of terrorism	• Force multiplier for kinetic activity • Intelligence preparation of the battlefield • Cripple opponent (sow chaos in society) • Psychological operations • May be ideological	States Nonstate actors	States including military units (cybercommand), nonstate actors, contractors

case of the 2016 US presidential elections, activist hacks which target the integrity of a political process may be as serious a threat to a nation as a traditional military cyber strike. When actors, including activist groups or other nations, interfere in domestic political events in another country, it may go beyond simple activism to constituting a crime according to international law.

The Professionalization of Hacking

Today, groups like the Association of Computer Machinery have formally drawn up codes of ethics and professional standards based on a shared understanding of computer science and the computing project. Such standards help practitioners understand their role in this project, the mission of computer science, and the standards of behavior and values that practitioners should adopt to participate in this project.

As we have noted, over time, computer security has become an established profession with its professional organizations and licensing mechanisms. Today, we can also point to the professionalization of principles for hackers, including the possibility of becoming licensed as a Certified Ethical Hacker. An *ethical hacker* or *white hat hacker* works for or with a private corporation or government agency to test their system's security. They may run tests – including conducting attempts to penetrate the system or using social manipulation tactics like phishing – to identify system vulnerabilities. They then make recommendations to the client regarding how to address these vulnerabilities. The ethical hacker thus doesn't seek to harm a corporation, but rather impersonates a hacker (thinking like the enemy) to help the corporation better protect itself.

Today, over 50,000 individuals have received the Certification in Ethical Hacking from the International Council of E-Commerce Consultants (EC-Council). The certification is endorsed by the US National Security Agency and the Department of Defense, and individuals throughout the world have taken the test to achieve the certification. In addition, the Certified Ethical Hacker Code of Ethics requires that certificate holders agree not to participate in any underground hacking community activities which involve black hat activities and not to participate or associate with black hat community activities that endanger networks (CEH Candidate Handbook v2.0). In addition to agreeing to work ethically for those who might employ them, white hat hackers often also contribute to the community of computer science professionals. Many participate in so-called *bug bounty programs*, reporting on security vulnerabilities that they have found incorporate government computer systems.

Box 3.2 Going Deeper: Ethics of Vigilantism

What Is Vigilantism?

A vigilante acts on their own accord (rather than as part of an official military or law enforcement organization) to address a perceived violation of the law. Vigilantes act in a premeditated rather than spontaneous fashion and employ violence or the threat of violence. Vigilantes hope to punish wrongdoers and help those in society feel that lawbreaking is being addressed (Dumsday, 2010).

Digital vigilantism is "a process where citizens are collectively offended by other citizen activity and coordinate retaliation on mobile devices and social platforms." Citizens may mobilize to punish acts like breaching social protocols, participating in terrorism, or participating in riots (Trottier, 2017).

Citizens who engage in online vigilantism may engage in acts like "naming and shaming," where, for instance, someone misbehaves in a public situation (for example, abusing a child, using hate speech, or engaging in violence), the video of their doing so is quickly uploaded and shared online, and online viewers quickly identify the perpetrator. In recent years, people have been fired from jobs or experienced other negative consequences due to being named and shamed online. Vigilantes may also engage in doxing – by making personal details about the person available online.

Recent Acts of Online Vigilantism – Factual and Fictional

In winter 2021, US citizens responded to the attempted insurrection at the US Capitol on January 6 by volunteering to assist law enforcement with identifying people who may have participated in these unlawful acts. Citizens used facial recognition software and other online detective methods to match photos to individuals and created databases to scan, view, and identify the images.

Is Vigilantism Ever Ethical?

However, most ethicists regard vigilantism as unethical – for several reasons. First, they worry that allowing individuals to enforce the law will, over time, lead to a decrease in citizens' respect for the authority of official law enforcement bodies like the police and the military. And, they argue, if people regard the state as less effective in pursuing justice than such private actors, then people may be less supportive of their government or democracy.

Ethicists also raise practical considerations. First, a vigilante "mob" can get out of control and make a situation more chaotic and violent.

In addition, a vigilante group may be more interested in pursuing its own political, economic, or social goals than providing the public good of law and order. In this way, vigilantism is not a benevolent or charitable activity aimed at creating a **public good**, but rather a selfish act to pursue a specific political, social, or economic end. For example, private immigration enforcers may have a less welcoming attitude towards immigrants to the United States than other citizens do. Therefore, their primary aim may be to discourage would-be immigrants to the United States from attempting to enter – rather than merely to protect US borders.

In considering online vigilantism, Trottier and others argue that "doxing" someone may constitute a breach of a person's privacy, which is often unlawful. Furthermore, they describe "naming and shaming" as unethical since the punishment or retaliation is not proportionate to the crime or act committed. Someone named and shamed online may have this act follow them for the rest of their lives. It may be the first information someone learns about them when looking them up online. It can prevent them from securing employment or forming relationships in the future.

Furthermore, online vigilantes may not have the same training as official law enforcement agents and may therefore be more prone to make errors in identifying perpetrators. Someone mistakenly identified as a terrorist, insurrectionist, or perpetrator of hate crimes may have their lives irrevocably altered, even if they are innocent.

For this reason, most ethicists reject the utilitarian argument in favor of online vigilantism.

Bibliography

Dumsday, T. (2010). On cheering Charles Bronson: The ethics of vigilantism. *Southern Journal of Philosophy*, *47*(1), 49–67.

Trottier, D. (2017). Digital vigilantism as weaponisation of visibility. *Philosophy & Technology*, *16*(2), 55–72.

Similarly, students must agree to adhere to the ISC Code of Ethics to receive Certified Information Systems Security Professional (CISSP) certification. This code includes two rules that are relevant here: A systems security professional should "protect society, the common good, necessary public trust and confidence, and the infrastructure" and "act honorably, honesty, justly, responsibly and legally" (International Information Systems Security Certification Consortium, Inc., 2022, p. 1).

Other ethical codes of conduct for computer professionals which address hacking behaviors include the ACM's ethical code, which also refers to "causing no harm." In addition, the ISC provides a code of ethics, as does the Information Systems Audit and

Control Association (ISACA). Finally, the Information Systems Security Association (ISSA) promotes a code of ethics similar to the ISC, ISACA, and ACM codes of ethics.

Of course, some skeptics like to point out that Edward Snowden, the individual who exposed the surveillance techniques of the National Security Agency, is also a Certified Ethical Hacker. Any of the techniques taught in the course can be used both in service of and against one's state or employer. Ultimately, it is up to the individual to use the skills developed in the course appropriately and in an ethical manner.

White Hat, Black Hat, and Grey Hat Hackers

In contrast to white hat hackers, *black hat hackers* attempt to breach internet security and gain unauthorized access to a system. They seek to destroy or harm the systems they penetrate, often by releasing viruses or destroying files. A black hat hacker could also engage in *cyber hostage-taking* by holding files for ransom or accessing personal information that the individual considered confidential (including educational information like test scores; or medical information) and release it or threaten to do so. Thus, black hat hacking could harm a specific individual, a class of individuals (such as hospital patients), or a corporation, agency, or nation. Black hat hackers' activities are frequently illegal, and they may work on their own or in collaboration with a criminal organization.

Box 3.3 Going Deeper: Ransomware

Ransomware is malware that can be secretly loaded onto a computer. It can encrypt all of a user's files. Once that is done, a message alerts the user that unless they pay a "ransom" (usually in cryptocurrency) to a specific address by a particular deadline, all of their data will be destroyed. Those who pay the ransom are given a key to decrypt the data. A new variant, the Chimera Crypto-Ransomware acts slightly differently; this ransomware encrypts your files and then delivers a demand for payment. Otherwise, your data will be released to the internet. In this way, this particular type of malware creates a threat not to destroy data but to embarrass the recipient of the malware publicly (Gamer, 2016).

Targets include commercial enterprises and parts of a nation's **critical infrastructure** – including schools, hospitals, and military assets. Attackers can include criminal enterprises, individuals, and increasing hostile government actors. A ransomware attack can thus be a nuisance to one's business, or it can be an act of war.

Ransomware attacks may be particularly unethical when used against those most vulnerable in our societies, such as those who are ill and receiving hospital treatment. If health care providers cannot access their medical records, patients may miss out on medication doses or lifesaving treatments. Ransomware attacks on hospitals increased during the 2020–2022 COVID-19 pandemic, a situation that criminals sought to exploit.

Reacting to Ransomware

Ransomware presents several legal and ethical dilemmas. For law enforcement agencies, it is often difficult to identify the agent who sent the malware. It is unclear how the sender might be prosecuted, even if they could be identified and caught, since attacks often come from countries where different laws and procedures apply. Furthermore, paying a ransom or negotiating with a terrorist group is not advised and is illegal under US and international law.

For those who are attacked, a different ethical dilemma is presented. Should you ever pay a kidnapper a ransom? Currently, many targets have indeed "paid up." A British study suggests that 40 percent of the victims of the Cryptolocker malware attack in Britain paid up. And in October 2015, Joseph Bonavolonta, Assistant Special Agent in Charge of the Cyber and Counterintelligence Program in the Federal Bureau of Investigation's Boston office, was publicly criticized when he admitted that even the FBI sometimes advises clients to pay the ransom (Zorabedian, 2015).

Ethics of Paying a Ransom

Ethicists give several reasons why you should never pay a ransom. First, while an individual or corporation benefits by paying a ransom and avoiding consequences against themselves, they may harm others – since paying up may encourage a group to carry out more attacks in the future. And a group might use the proceeds from *your* ransom to carry out research and development activities to make better weapons in the future. Paying your ransom and saving your data is thus viewed as selfish since it does nothing to protect others from future attacks and perhaps even makes them more likely (Zorabedian, 2015).

However, McLachlan (2014) argues that people have a "duty of care" that requires them to consider the needs of their dependents over a more general duty to all of humanity. One could argue that the hospital's duty is to its patients in cyberspace. Allowing their medical duty to be destroyed would breach the relationship between the hospital and its patients,

There appears to be no clear ethical consensus, and perhaps no good solution in this area.

Bibliography

Gamer, N. (2016, January 8). Ransomware one of the biggest threats in 2016. *TrendMicro Simply Security*. Retrieved January 15, 2016, from http://blog.trendmicro.com/ransomware-one-of-the-biggest-threats-in-2016.

Information Security Institute. (2016). *ISC code of ethics*. Retrieved June 1, 2016, from www.ISC2.org/ethics.

McLachlan, H. (2014, October 3). Paying ransom for hostages is sometimes the right thing to do – here's why. *The Conversation*. Retrieved October 1, 2016, from http://theconversation.com/paying-ransom-for-hostages-is-sometimes-the-right-thing-to-do-heres-why-32460.

Riggi, J. (n.d.). *Ransomware attacks on hospitals have changed*. Retrieved from AHA.org.

Singer, P. (2014). The ransom dilemma. *Project Syndicate*. Retrieved June 1, 2016, from www.project-syndicate.org/commentary/islamic-state-hostages-ransom-by-peter-singer-2014-12.

Zorabedian, J. (2015). Did the FBI really say "pay up" for ransomware? Here's what to do . . . *Naked Security*. Retrieved June 1, 2015, from http://nakedsecurity.sophos.com/2015/10/20/did-the-fbi-really-say-pay.

In contrast, white hat hackers receive training. They agree to adhere to explicit rules regarding who they will target in conducting penetration tests of a system, the sorts of warnings they will provide to the target before commencing their attack, and the responsibility they accept in conducting the attack. In addition, white hat hackers seek and are granted permission before attempting to access a system, place self-imposed limits on the damage they will create in conducting attacks, and keep detailed logs of the activities they engage in while attempting to "own" or "pwn" a system, and conduct legal tests of a system which are spelled out in an explicit contract before attempts are made.

Both white hat and black hat attackers thus use similar techniques in attempting to access and own a system, but they do so under very different ground rules. Table 3.2 compares and contrasts the two sets of assumptions under which each group operates.

Table 3.2 White Hat Versus Black Hat Hacking Activities

	White Hat	*Black Hat*
Target	Ground rules are established with defender regarding targets, what is off-limits	May choose a target as the result of a grudge or for profit; no mutual decision with the defender about what targets are off-limits
Warning?	Attacker and defender agree on when an attack will begin; employees may be warned	No warning that an attack is imminent
Responsibility for Attack	The attacker is known to the defender and takes responsibility for the attack.	The attacker is not known to the defender, and the defender may be unable to trace back due to the attribution problem
Access	The defender is provided with an invitation to attempt to gain access	Access is illegal, represents a trespass or invasion
Damages	The attacker takes trophies, i.e., compromising information, only to raise defender awareness of vulnerabilities; treats compromising information carefully and with respect; takes steps to preserve access to information by the defender so it can be restored	Trophies may be taken and used for profit or public embarrassment; information or systems may be permanently destroyed
Intent	The attacker seeks to own the system to demonstrate to the defender that it is possible	The attacker seeks to own the system to exploit it for economic, political, or ideological purposes
Record of Activity? **Legal?**	The attacker does not cover tracks by destroying logs so that defender can learn what transpired and take steps to fix or correct access problems	
Legal?	Penetration testing is legal if it is spelled out in a contract between the defender and attacker	May violate the United States Criminal Code statutes on fraud wire and electronic communications interception; may violate the Digital Millennium Copyright Act; may violate the Cybersecurity Enhancement Act of 2002

Source: Harper et al. (2011, pp. 8–9) provided the basis for reasoning in this chart.

In identifying a hacker as a white hat or black hat hacker, then, the most crucial question is whether or not they gained unauthorized access to a system, rather than the type or amount of damage sustained; whether the damage was intentional or unintentional; or who specifically they were working for. (However, this situation is complicated during activities like warfare, since one might legitimately claim to be acting ethically even if they are "cracking" a system – if they are doing so on behalf of their state against an enemy state in wartime, or if they are acting on behalf of an intelligence agency. We will explore this issue more in our chapter on cyberwarfare.)

The final type of hacker to be considered is the grey hat hacker (Harper et al., 2011, p. 18). In essence, a grey hat hacker is self-employed, working to collect bug bounties through testing systems without authorization, yet seeking not to damage the systems but rather to enrich himself by collecting rewards for identifying system vulnerabilities. This behavior can be justified on ethical grounds by arguing that hacking produces a public good by identifying security vulnerabilities.

State-Sponsored Hacking

States have utilized hacking as a weapon of war since 2010 when the United States and Israel combined forces to hack into a facility in Iran that they believed was being used to enrich uranium, a component of nuclear weapons. They were able to send malicious code (known as Stuxnet) to the facility, degrading the centrifuges used in the uranium enrichment process (Hernandez, 2021).

Today, states may engage in direct hacking (by a group like the US military's Cyber Command) or indirectly – through outsourcing activities to a third party, including private contractors. As a result, both domestic and international law enforcement agencies struggle with the problem of attribution or assigning blame to a particular actor for the damage caused. In addition, such actions may create legal ambiguities – since they often traverse the lines between criminal activities, acts of war, and violations of international law.

Law enforcement agencies can track state-sponsored hacking by looking at the level of sophistication involved in such attacks. Advanced Persistent Threat (APT) refers to ongoing, highly sophisticated hacking activities by a well-resourced state-level actor. APT often includes activities in which an actor accesses a system and remains undetected within that system for a prolonged time. APT hacking activities may include the placement of zero-day exploits in a system, which can then be "detonated" remotely in response to a specific event or command. Other APT hacking activities include attempts to map a system, including its vulnerabilities. The hack may not involve an attack on the system itself in these situations. Still, it may instead constitute "intelligence preparation of the battlefield" in preparation for a later attack on the system or its owners.

Analysts disagree on whether APT hacks should be considered an act of war. Some analysts argue that they are merely a new and novel way for actors like the intelligence community to engage in covert (or undeclared) warfare against an enemy.

The existence of such actors constitutes part of today's threat landscape. It seems unlikely that such groups will disappear because today's threat attack surface is so great, with many targets existing on all types of machines, from laptops, to social media sites, to Internet of Things devices. States will continue to utilize this type of warfare due to the low cost and ease of use.

Box 3.4 Tech Talk: Cryptohacking and Cryptojacking

Ever since the advent of the world's first cryptocurrency in 2006, individuals and corporations have begun moving towards using cryptocurrency for carrying out financial transactions and saving and investing. You may be familiar with some cryptocurrencies, including Bitcoin, Ethereum, and Dogecoin.

What Is Cryptocurrency?

But what is a cryptocurrency? Cryptocurrencies are "coins" that are "mined" by computers. A crypto miner can earn a coin by being the first to solve a calculation that requires a great deal of computing power. Coins can also be purchased.

Cryptocurrency exists only virtually. There is no physical paper money. Instead, transactions are tracked electronically using Distributed Ledger Technology. Transactions are recorded, and the records are updated automatically on all of the computers which form part of the network. In this way, a chain of custody is created which tracks cryptocurrencies like Bitcoin (Malwarebytes.com, n.d.).

Cryptocurrency thus differs from conventional currency since a central bank does not issue it, and since nations do not have control over cryptocurrency monetary policy – as they do with conventional currency. Furthermore, in a conventional banking scenario, particularly after the terrorist events of 9/11, anonymous financial transactions are mostly disallowed. In contrast, people often conduct transactions using virtual cryptocurrencies anonymously.

Thus, cryptocurrency transactions may be particularly enticing for people doing illegal things (such as not paying taxes on their income, or making money by doing something illegal like selling drugs).

How to Hack a Cryptocurrency

So how does one steal cryptocurrency? Cryptocurrency hacks often involve phishing scenarios, where an individual clicks on a link that either infects their computer or loads a crypto mining code onto their computer. In this way, the actor "steals" computing power from a victim (or cryptojacks it), using that corporation's computer power and space to mine cryptocurrency for their use. Such attacks can significantly slow down a system and, in some cases, will damage a system's computers as the crypto mining program deletes or moves files to create more space for crypto mining. Such attacks are not very technologically sophisticated. Indeed, a crypto mining kit can be purchased online for about $30, so the barriers to entry for this type of crime are relatively low. There is also at present a relatively low risk that the perpetrator will be caught.

Why It Matters

In 2021, hackers were estimated to have stolen approximately $14 billion worth of cryptocurrency using the attacks described here and more conventional scamming attacks. A hacker might also breach a blockchain information wallet and change the ownership records for someone's cryptocurrency. (Such attacks on blockchain information may also be part of a larger money-laundering scheme.) Some authors, like Smiljanic Stasha, argue that blockchain accounting technology is highly vulnerable to these types of attacks. She worries that if quantum computing matures too quickly, similar issues might arise with this new emerging technology.

Bibliography

Malwarebytes.com. (n.d.). Cryptojacking – what is it? *malwarebytes.com*. Retrieved June 1, 2022, from www.malwarebytes.com/cryptojacking.

Stasha, S. (2022, February 13). Cryptocurrency hacking statistics. *policyadvice.net*. Retrieved June 1, 2022, from https://policyadvice.net/money/insights/cryptocurrency-hacking-statistics/.

Ethics of Pen Testing

Penetration testing, or *pen testing*, has been described as "the (sanctioned) illegitimate acquisition of legitimate authority" (Pierce et al., 2006). Pen testing is a set of practices usually carried out by an outside company hired by a corporation to attempt to access their systems. Pen testers utilize traditional hacking and social engineering techniques to break into the client's system to identify and aid in fixing any security vulnerabilities that are found. *Social engineering* is defined as "the practice of obtaining computer information by manipulating legitimate users" (Gupta, 2008, p. 482). It is a form of deception in which testers might impersonate legitimate users and make inquiries about passwords or send e-mails inviting users to click on a link. It relies less on technology and more on the human element to access a system (Allen, 2006). Because social engineering involves deception, it raises ethical issues.

Despite the ethical issues, pen testing effectively increases system security and is widely used today. A recent survey indicated that 34 percent of companies conduct external penetration tests, and 41 percent conduct internal penetration tests (Trustwave, 2016). Penetration testing is the most widely outsourced security activity among corporations. Indeed, in April 2016, the Pentagon, headquarters of the US Department of Defense, launched its Hackathon, encouraging users to attempt to access the over 200 sites associated with the Department of Defense (Krishnan, 2016).

But penetration testing practices contain potential legal and ethical pitfalls. How can corporations and personnel conduct activities that are lawful and ethical?

Ensuring the Legality of Pen Testing

Currently, analysts recommend that all pen testing parties create well-documented, written agreements. Many companies specify that those hired to carry out penetration testing have specific certifications (Certified Ethical Hacker; Information Assurance Certification Review Board Certified Penetration Tester; or the Council of Registered Ethical Security Testers certification [CREST]). Testers often undergo specific ethics training and furnish evidence that they have a clean criminal record of getting certified. Testers often also sign a non-disclosure agreement (NDA) spelling out their responsibilities to protect company secrets and their agreement to comply with the country's laws where work is being done. It is still illegal for pen testers to impersonate law enforcement, threaten to harm someone to get information, or obtain federal documents to access Social Security numbers. Companies will often ask testers to sign specific contracts spelling out what sorts of behaviors are authorized. For example, suppose testers go through dumpsters looking for information that might aid in penetrating a system. In that case, they should have permission to do so for incidents like stealing an employee's computer.

Ethics Models

However, even if you follow all the rules, ethics issues will still arise. We can apply the three lenses – virtue ethics, consequentialism, and deontological ethics – in considering these issues.

Virtue Ethics

A virtue ethics approach rests on the assumption that one's actions are a function of one's character. It does not allow for a division of actions into, for example, those carried out in public life and those carried out in private life. It also does not focus merely on the outcome of actions but asks the actor to consider what motivates his action and what that says about his character.

Pen testing presents an ethical issue from a virtue ethics approach for three reasons:

- A pen tester employs deceptive practices ("lies"), including misrepresenting himself and his desires to achieve his target.
- A pen tester often adopts an instrumental approach towards other human beings, violating the principle of respect for others. To access data, the pen tester may treat the target group not as individuals worthy of inherent respect and dignity, but rather as sources of information the pen tester needs to access.
- A pen tester may display behaviors that don't seem morally consistent – for example, allowing deception in carrying out social engineering practices, but not elsewhere in a person's life.

Self-restraint and empathy are two virtues that can guide the pen tester who wishes to behave ethically.

The Virtue of Self-Restraint

There are many examples of individuals who have cultivated the discipline of self-restraint as a virtue in history. A hacker who cultivated restraint would accept that opening the door to a system does not necessarily mean they have the right to enter. Instead, when entering that system would be destructive or harmful, the hacker might decide to forgo this opportunity and its rewards.

The Virtue of Empathy

The next virtue that an ethical hacker might cultivate is empathy or respect for others. Here Brey (2007) argues that hackers commit an ethical breach when they commit activities that compromise the ability of individuals to control the confidentiality, integrity, and availability of their data (p. 23). A hacker motivated by respect might thus decide not to release information like photographs or videotapes if doing so would hurt an individual.

So how might a virtuous pen tester behave? A virtue ethics approach thus suggests three cautions:

First, the tester should treat everyone with respect in carrying out tests, not harming or embarrassing employees, but instead reporting the results of social engineering tests anonymously. Pierce et al. (2006) suggest reporting only percentages and statistical information (for example, noting that half of the employees clicked on the link containing malware rather than naming specific people who did so). An employee might be embarrassed or fired if linked to specific activities.

Second, the tester should include a representative of the Office of Human Resources in initial contract meetings. That representative should avoid engaging in scenarios that employees would find upsetting or in actions that cause employees embarrassment or jeopardize their employment or reputation. Goodchild cautions against impersonating an actual employee of the company when sending an e-mail that might get an employee to share a file, information, or data. Instead, she recommends creating a fictional employee. She notes that it might be acceptable to lie and say you left your keys on your desk to get readmitted to a building, but you should not make up a story about a car accident that might traumatize some people. Pierce et al. also raise ethical concerns regarding the marketing of penetration testing services. They argue that sowing fear, uncertainty, and doubt (FUD) to sell ethical hacking services is unethical, as is the use of crime statistics to promote these services. They suggest that pen testing companies must walk a fine line between informing potential consumers of genuine risks their companies may face and needlessly scaring or worrying potential clients (Pierce et al., 2006, p. 198).

Finally, pen testers can practice self-restraint and refrain from accessing information they might find in a system (like personal records or e-mail) if he is not required to do so as part of their contract (Faily, 2015, p. 239).

Box 3.5 Critical Issues: Can You Hack a Voting Machine?

How easy is it for a hacker to breach a voting machine, and why does it matter? At a 2019 meeting of the DefCon Cybersecurity Conference titled "Voting Village," participants were challenged to see if they could hack into any of the most widely used voting machines in the United States. Many participants purchased their voting machines through eBay's online auction site and found it relatively easy to guess the passwords on these machines. From there, they proceeded to experiment – and found that most machines had relatively weak encryption protocols. Further, they found it easy to erase data, change data, and shut down machines. Describing the conference results, Marks notes that if ethical hackers could breach all voting machines (and purchase voting machines from eBay), we might assume that adversary nations have the same capabilities (Marks, 2019).

In addition to the threats posed by unsecured machines, the companies that manufacture voting machines can also be a target of hacker activities, from simple to complex. In

2018, CSO Online found that many voting machine manufacturers in the United States had data that had been compromised or leaked, available on multiple channels such as the Dark Web. The information included employee passwords and even employee e-mails.

CSO also noted that many people in the United States who volunteer as election personnel are older, retired, and less familiar with the technology. Many voting machine companies, therefore, still offer remote access to machines. This means that if a problem with a machine develops, the manufacturer might log into that machine to attempt to fix the problem (Porup, 2018). Despite the expert recommendation that voting machines should be "air gapped," or not connected to the internet, it appears that many machines still are not.

Why Having a Safe Voting System Matters

One of the factors that makes a democracy a democracy is that democracies have "regular elections and a peaceful transfer of power." In other words, when an election occurs (usually every two or four years in the United States), people can accept the election results as just and fair because they have confidence in how the election was conducted. As a result, elected officials can accede to power, and those who are not elected must step down.

If people have questions about the legitimacy of an election or how it was carried out, they may be less likely to vote in future elections. They may have less trust in government or desire to participate in a democratic society. Furthermore, they may have less trust in their elected leaders if they don't trust the process of choosing those leaders.

Finally, if an election system is not secure, outside nations might interfere in the outcome of an election, overriding the citizens' choices. Suppose a foreign power attempts to alter or change an election result. In that case, this constitutes a breach of a nation's security, since international law requires that the citizens of a country be the only ones able to alter events within a nation.

The US Government Response

Thus, nations take guaranteeing the security of their elections very seriously. In 2018, the US government formally acknowledged that voting machines constitute critical infrastructure for national security. That same year, the US Cybersecurity and Infrastructure Security Agency (CISA) was established under the US Department of Homeland Security. However, this agency has been criticized for having few tools to respond to breaches and enforce critical infrastructure laws. Although CISA has responsibility for election security, so far it has been most active in providing training and coordination between election agencies on the local, state, and national levels through the Elections Infrastructure Information Sharing and Analysis Center (EI-ISAC).

Should an Ethical Hacker Ever Hack an Election?

In this chapter, we have looked at various types of hackers, including nuisance hackers who might attempt to access a target just for the thrill of it to see if they can. However, the ACM Code of Ethics cautions computer experts to be aware of and respect a nation's laws. In this author's opinion, the risks associated with election hacking, including the risk to democracy itself, are too significant for it ever to be a viable option for an ethical hacker. The only exceptions to this rule would be if the hacker participated in a bug bounty or similar challenge where the goal was to hack machinery to point out flaws to make future systems more secure.

Bibliography

Marks, J. (2019, September 27). The cybersecurity 202: US voting machines vulnerable to hacks in 2020, researchers find. *Washington Post*. Retrieved June 1, 2022, from https://democracychronicles.org/are-voting-machines-too-vulnerable-to-hacking/.

Porup, J. M. (2018, March 30). Want to hack a voting machine? Hack the voting machine vendor first. *CSO*. Retrieved June 1, 2022, from www.csoonline.com/article/3267625/want-to-hack-a-voting-machine-hack-the-voting-machine-vendor-first.html.

The Problem of Competing Values

Throughout this book, we have emphasized the ethical dilemmas that can arise when a computer scientist must decide between several solutions to a problem, each based on a different value prioritization or where competing values come into play. In thinking about identifying vulnerabilities in a computer system through penetration testing, computer programmers may again encounter the problem of competing values and the need to make trade-offs between them.

One value that a corporation might prioritize is efficiency, or the ability to solve a problem as quickly as possible (which certainly matters to product consumers!). Thus, a company's leadership may disagree with your recommendation as a pen tester regarding the need to implement multiple-factor authentication for all employees or restrict administrator access to only a few individuals. They might express concerns that making a system more secure might also slow down their ability to operate on it or solve problems quickly. How might you speak back to that concern?

Another value that corporations often prize is equity or access. Imagine a situation where, for example, a company creates a classified report that they wish to circulate to only a tiny circle of company leadership. However, the company's leadership includes an individual who is visually impaired. This individual needs to use a program that "translates" written graphics and charts into haptic (or touch) products so that she can interact with the charts. This translation service is outside the company's system and requires that charts be uploaded to an outside client to be accessible to individuals with disabilities. How might you best protect the confidentiality of all information while simultaneously honoring the company's desire to provide equity and access to all of its employees? How do you weigh disclosure risks against the need to provide equity and access?

Consequentialism

In contrast to the virtue ethics approach, which considers what decisions about how to behave indicate the hacker's character, a utilitarian ethics approach would consider only the consequences of the hacking offense. Who was hurt by the hacking, and what sorts of damages occurred? This approach assumes that not all types of hacking are alike. Some are ethical, while some are not. It also allows for a spectrum of ethicality, with actions being somewhat unethical, primarily ethical, or completely unethical.

Here, we can identify valuable arguments both supporting and condemning hacking. Defenders believe that hacking produces social benefits which outweigh the costs and inconvenience that targets might incur (Raicu, 2012). They suggest that hacking increases transparency and accountability within the system of government and corporate actors using computers. However, as Raicu asks, if the outcome associated with hacking is that the scales of justice will somehow be righted and social wrongs will be rectified, is the hacker necessarily the best individual to be carrying out this righting of the scales – rather than some other type of regulatory actor, such as the government or a nongovernmental organization?

Not surprisingly, consequentialism favors pen testing and the attendant social engineering. The goods achieved through pen testing include the ability to gain awareness of weaknesses and vulnerabilities and an increased capability to address these vulnerabilities (Pierce et al., 2006). The only qualm a consequentialist might feel is due to an ethical concern that Pierce et al. raise. They note that some corporations may be too reassured if pen testing fails to find vulnerabilities in their companies. They argue that the failure to find vulnerabilities doesn't necessarily mean that there weren't any, merely that none were found. They thus warn that pen testing might provide false reassurance, ultimately harming a company by making it feel secure although it is vulnerable.

Deontology

Radziwill et al. note that deontological ethics would most likely not support hacking, regardless of the situation's specifics. They note that the doctrine of reversibility would require a moral decision-maker to ask, "How would I feel if my site were hacked?" However, they then state that since white hat hacking, including pointing out the flaws of one's system, *would* be appreciated, it may be morally justifiable to help others out in the same way (Radziwill et al., 2016).

In considering penetration testing, a deontologist would start with the **categorical imperative**, asking if there is a principle that guides penetration testing, which would still be valid were it to become a universal rule. It is difficult to see how support for either deception or other social engineering activities could be upheld as a universal rule. Indeed, if one accepts the contention that lying to an employee for reasons of achieving a password or entry into a computer system is okay, then one would also need to accept that lying to an employee for other reasons at other times would also be morally acceptable – since the categorical imperative rule leaves no place for the application of detailed ethics. Applying deontological ethics would also require the pen tester to accept that he would be okay with any activities he might carry out upon another person to gain entry into a system. In other words, if he found it morally acceptable to impersonate a fellow employee or supervisor to manipulate an employee into revealing a password, he should also be accepting of a colleague or employer perpetrating the same hoax upon him.

A deontologist might also object to the fact that pen testing often requires an instrumental approach towards other human beings, violating the principle of respect for others. To get access to data, the tester may treat his target group not as individuals worthy of inherent respect and dignity, but rather as sources of information that he needs to access. Indeed, Bok (1978) notes that lying to coerce someone to do something (such as giving up information) represents a form of violence against that subject.

However, suppose one decided to go ahead with pen testing anyway. In that case, a deontological ethics approach suggests three cautions: First, the tester should treat everyone with respect in carrying out tests, not harming or embarrassing employees but instead reporting the results of social engineering tests anonymously. Pierce et al. (2006) suggest reporting only percentages and statistical information (for example, noting that half of the employees clicked on the link containing malware rather than naming specific people who did so). An employee might be embarrassed or fired if linked to specific activities.

As this chapter has shown, the line between unethical and ethical hacking is not always clear. Many ethical grey zones exist as technology quickly adapts and changes, often outpacing our ability to regulate it and establish a consensus on ethical issues. In some instances, expediency (like the need to distribute information about vaccine availability as quickly as possible) may seem more important than the need to respect intellectual property or the laws of ownership. But ethical frameworks like utilitarianism allow us to consider harms and benefits, while ethical codes enable us to consider specific issues within a larger context.

CHAPTER SUMMARY

- As cybersecurity has evolved as a profession, hacking has moved from a casual, unregulated activity to one with clearly stated norms and standards of professional practice.
- Today, we can distinguish between white hat and black hat hackers, with white hats often assisting government and groups in improving their security through running pen tests.
- Many types of hacks are illegal. The distinction between cyberterrorism and hacking is not always clear.
- Both technology developers and users share ethical responsibilities. Thus, it is possible to identify a duty not to hack and safeguard one's material so that one is not hacked.

DISCUSSION QUESTIONS

1 Anonymous and the Ethics of Vigilantism

In this chapter, we have discussed hacking for good and malicious hacking. We have also looked briefly at the ethics of vigilantism.

Please read this article about the international collective actor Anonymous and how it has participated in the Russia–Ukraine conflict. The article can be found at: www.theverge.com/2022/3/11/22968049/anonymous-hacks-ukraine-russia-cybercrime-danger

In spring 2022, Anonymous and its supporters: hacked into Russian television stations, disrupting the airwaves with news about the war; sent over 700,000 individual text messages to Russian citizens telling them about the war; hacked into Russia's communications satellites; and hacked into Russia's Ministry of Defense, where they released e-mail addresses of high-ranked defense department officials to the public.

- Use the three paradigms to consider whether or not these actions could be justified as ethical. What values or virtues might Anonymous claim to represent? What might a deontologist think about such actions? (Here, you might consider the statements made by a blogger, who stated that "Like it or not, these actions violate national sovereignty. How would you like it if during the next election in your country, Russian activists began messaging your nation's citizens, telling them how to vote?")
- Do you regard these actions as moral vigilantism or unlawful activism?

2 Is Doxing Ever Ethical?

Read the following article on the ethics of doxing: www.luc.edu/digitalethics/researchinitiatives/essays/archive/2017/theethicsofdoxingnazis/

In this article, the authors suggest that **doxing**, or publicly identifying, individuals who have participated in activities like attending a Nazi rally in the United States can be moral.

- Identify the ethical framework (utilitarian, deontological, or virtue ethics) used in this article and explain its application. Then, see if you can make an ethical argument in favor of doxing using either of the other two perspectives.
- Finally, attempt to do the reverse: What would a utilitarian argument against doxing look like? A virtue ethics argument? A deontological argument?
- Would you ever participate in an effort like this? Why or why not?

RECOMMENDED RESOURCES

CNBC. (2021, January 22). The Solar Winds hack and the future of cyber espionage. (10 minute video). Retrieved June 1, 2022, from www.youtube.com/watch?v=jxTxGlE9X5s.

Pierluigi, P. (2013, September 18). Hacking satellites . . . Look up to the sky. *Information Security Institute*. Retrieved from https://resources.infosecinstitute.com/topic/hacking-satellite-look-up-to-the-sky/

Riggi, J. (2021). Ransomware attacks on hospitals have changed. *American Hospital Association*. Retrieved June 1, 2022, from www.aha.org/center/cybersecurity-and-risk-advisory-services/ransomware-attacks-hospitals-have-change www.aha.org/center/cybersecurity-and-risk-advisory-services/ransomware-attacks-hospitals-have-changed–:~:text=By%20John%20Riggi%2C%20Senior%20Advisor%20for%20Cybersecurity%20and,patient%20care%2C%20which%20puts%20patient%20safety%20at%20risk.

Schneier, B. (2021). *The coming AI hackers*. Cambridge, MA: Belfer Center. Retrieved June 1, 2022, from Available at www.belfercenter.org/publication/coming-ai-hackers.

References

Allen, M. (2006). *Social engineering: A means to violate a computer system*. SANS Institute Reading Room. Retrieved December 1, 2015, from www.sans.org/reading-room/whitepapers/.

Bok, S. (1978). *Lying: Moral choice in public and private life*. New York: Pantheon Books.

Bratus, S., Shubina, A., & Locasto, M. E. (2010). Teaching the principles of the hacker curriculum to undergraduates. In *Association for Computing Machine Special Interest Group on Computer Science Education (SIGCSE)*. Milwaukee, WI: Association for Computing Machinery.

Brey, P. (2007). Ethical aspects of information security and privacy. In M. Petkovic & W. Jonker (Eds.), *Security, privacy and trust in modern data management* (pp. 21–36). Heidelberg; Berlin: Springer.

Cox, J. (2021, August 5). Scammer service will ban anyone from Instagram for sixty dollars. *Motherboard*. Retrieved June 1, 2022, from www.vice.com/amp/en/article/k78kmv/instagram-ban-restore-service-scam.

Faily, S. M. (2015). Ethical dilemmas and dimensions in penetration testing. In *Proceedings of the ninth international symposium on human aspects of information security and assurance (NAISA)*. Morrisville, NC: lulu.com.

Fisler, C., Beard, N., & Keegan, B. C. (2020). *No robots, spiders or scrapers: Legal and ethical regulation of data collection methods in social media terms of service*. Palo Alto, CA: Association for the Advancement of Artificial Intelligence.

Gupta, J. N. D. (2008). *Handbook of research on information security and assurance*. New York: IGI Global.

Harper, A., Harris, S., Ness, J., Eagle, C., Lenkey, G., & Williams, T. (2011). *Grey hat hacking: The ethical hacker's handbook*. New York: McGraw-Hill Osborne Media.

Hernandez, L. (2021, October 3). What are state-sponsored cyberattacks? *F-Secure Blog*. Retrieved June 1, 2022, from https://blog.f-secure.com/what-are-state-sponsored-cyberattacks/.

International Information System Security Certification Consortium. (2022). *Code of ethics*. Alexandria, VA. Retrieved from http://www.isc2org/ethics.

Krishnan, R. (2016, March 3). Hack the pentagon – United States Government challenges hackers to break the security. *Hacker News*. https://www.usds.gov/projects/hack-the-pentagon.

Pierce, J., Jones, A., & Warren, M. (2006). Penetration testing professional ethics: A conceptual model and taxonomy. *Australasian Journal of Information Systems*, *13*(2), 193–200.

Radziwill, N., Romano, J., Shorter, D., & Benton, B. (2016). The ethics of hacking: Should it be taught? *OALibrary Journal*. Retrieved October 21, 2016, from www.oalib.com/paper1405941.

Raicu, I. (2012). *Unavoidable ethical questions about hacking*. Santa Clara, CA: Markkula Center for Applied Ethics.

Reuters. (2021, August 1). Hackers shut down system for booking COVID-19 shots in Italy's Lazio Region. Retrieved June 1, 2022, from www.reuters.com/world/europe/hackers-shut-down-system-booking-covid-19-shots-italys-lazio-region-2021-08-01/.

Tavani, H. (2004). *Ethics and technology: Ethical issues in an age of information and communication technology*. Hoboken, NJ: Wiley Publishing.

Trustwave. (2016). *Security testing practices and priorities: An Osterman research survey report*. Retrieved from https://www.trustwave.com/en-us/resources/blogs/trustwave-blog/introducing-security-testing-practices-and-priorities-an-osterman-research-survey.

Xu, Z., Qing, H., & Zhang, C. (2013). Why computer talents become computer hackers. *Communications of the ACM*, *56*(4), 64–74. Retrieved October 21, 2016, from https://cacm.acm.org/magazines/2013/4/162513-why-computer-talents-become-computer-hackers/fulltext.

Young, R. Z., Zhang, L., & Prybutok, V. R. (2007). Hacking into the minds of hackers. *Information Systems Management*, *24*(4), 281–287.

Part II

4 The Problem of Privacy

LEARNING OBJECTIVES

At the end of this chapter, students will be able to:

1. Apply the three frameworks – virtue ethics, deontological ethics, and utilitarian ethics – to describe privacy ethics
2. Describe privacy as both a universal and relative concept
3. Articulate ongoing legal issues regarding regulating privacy in the United States
4. Describe competing values that may be important in thinking about both personal and data privacy

As we consider the ethical problem of privacy in this chapter, we can consider three real-life situations involving privacy rights in cyberspace:

- In 2020, a driver for the ride-sharing service Uber posted online that his company was able to make certain assumptions about the individuals who hailed the ride-sharing service. In particular, he noted that based on data they collect, Uber could assess which of their customers have had a "one-night stand" (a short-term sexual encounter with a relative stranger). As a result of these revelations, Uber was the subject of a lawsuit in New York City, and the platform changed its data sharing and storage practices. In particular, it limited the number of individuals permitted access to the database's so-called "God View," which allowed for unlimited viewing of all user activity. In addition, better practices were implemented for data anonymization, and more location data for Uber drivers and customers were encrypted.
- In 2019, American journalists uncovered a university yearbook photo from 1984, which purported to show Virginia governor Ralph Northam at a university party where he wore "black face makeup," masquerading as an African American man, although he was white. Some individuals called upon Northam to resign, believing that the photo showed that he had racist tendencies, which made him unfit to govern the racially diverse US state. Northam claimed that the photo was not him. In contrast, others claim that his decision to masquerade as an African American man was merely a "youthful indiscretion" that was no longer relevant to his present-day qualifications and responsibilities (Today.com, n.d.).
- In 2020, author Gregory Rasner described how the COVID pandemic had proven to be a "bonanza" for cybercriminals, both domestically in the United States and internationally. Rasner described how many government offices and commercial entities

DOI:10.4324/9781003248828-6

> attempted to pivot from carrying on their day-to-day operations in an office environment to working online quickly. He argued that initially, most companies lacked adequate information protection measures. Employees used home routers and less secure networks, vulnerabilities existed in Zoom and other online meeting software, and people routinely opened phishing e-mails, fearing they might be missing an important communication from their employers. In addition, some COVID information websites and COVID-related emails were used as vehicles to disseminate malware (Rasner, 2020).

All three of these examples illustrate that data practices in cyberspace often have unintended consequences. Data that might seem initially uncontroversial (such as location data for an Uber ride) can be combined in new and novel ways to yield new and potentially intrusive or harmful insights. Events that occurred many years ago may live on in cyberspace and can follow people throughout their lives. And in a crisis like a pandemic, privacy might seem like an afterthought with harmful consequences to individuals, corporations, and states.

But the examples also raise several ethical issues: What are the corporations' responsibilities that collect our data, and what are our responsibilities to practice good cyber hygiene and safeguard our data? Should everyone have an equal right to be provided with privacy regarding their online behavior, or should those whose actions are unethical, illegal, or criminal expect that they would have less privacy? Should the actions we carry out at work be regarded as automatically subject to our employer's surveillance and supervision?

In this chapter, we ask: what is privacy, and what rights do people have regarding the privacy of their activities, data, and digital identities? Are there universal rules regarding what it means to safeguard privacy, and how are recent technologies changing our orientations towards questions of privacy? Will privacy even exist in the future?

What Is Privacy?

"Privacy" is a complex phenomenon encompassing issues, including the integrity of our bodies, our reputations, and our data. Privacy is often described as a cluster of rights – including how we can represent ourselves to others, the control we have over how we are perceived, monitored, or represented by others, as well as the access that others have to our persons and information about ourselves, and the degree to which we can control other's control of this access to our persons and data. In a landmark 1890 essay entitled "The Right to Privacy," Samuel Warren and Louis Brandeis defined privacy as "a right to decide how you present yourself and how you are perceived" (Warren & Brandeis, 1890). In other words, privacy allows an individual to keep information or data secret or confidential and to decide with whom to share secrets and under what conditions. A *secret* refers to information not known or intended to be shared with others.

The term *privacy* thus includes rights to your own body, rights to your property, and rights to your information or data generated by or about you. Floridi (2013) suggests that we distinguish between physical privacy (the right not to be observed or to be "left alone"), decision privacy (the right to make decisions about your own life without interference by others), mental privacy (the ability to have our thoughts) and information privacy (the right to control what others know about you and what information is collected and shared about you) (Tavani, 2008).

Privacy means that you as an individual have a claim to own your body, your image, and any information you generate about yourself or any information generated about you. You are the person who controls that information – who gets to see it, who does not, and under what circumstances (Al-Saggaf, 2015). As the creator of that information and data, you get to decide who gets to be an intimate friend who knows everything about you, who gets to know some information about you, and who should not be permitted to see information and data about you (Marmor, 2015, p. 1).

Privacy also includes the idea of protection; Privacy protects us from "unwanted access by others – either physical access, personal information or attention" (Wilkins, 2008, pp. 10–11). Thus, privacy is related to *surveillance*. The right to privacy means that we have a right not to be surveilled – or watched or monitored – without our consent or knowledge, to be informed when we are under surveillance, and to be able to establish private places in our own lives where we will not be under surveillance either by other individuals or by a group or agency, such as the government.

Many professions have developed explicit ethical codes that spell out how information (or secrets) should be treated and the conditions under which such information should be shared or kept private. For example, the American Bar Association Model Rule 1.5 states that lawyers cannot reveal information that their clients have shared without clients giving their informed consent. This rule helps ensure that clients can trust those they have hired to represent them. Attorneys may violate that privilege only in rare circumstances (Michmerhuizen, 2007).

A violation of privacy occurs if "someone manipulates, without adequate justification, the environment in ways that significantly diminish your ability to control what aspects of yourself you reveal to others" (Marmor, 2015, p. 14). This may include, for example, a situation where the government tells citizens to assume they are being monitored all the time or a situation where a corporation changes people's privacy settings on social media without their authorization and consent.

Public Space and Private Space

Philosophers often distinguish between public and private space and public and private information. We can trace this idea back to Aristotle, who distinguished between the public sphere of governance (the *polis*) and the private sphere of the household (the *Oikos*). He argued that people developed themselves through having a private space for contemplation (Moore, 2005, pp. 16–17). Confucius also distinguished between the public activity of government and the private affairs of family life (Moor, 1997)

Legally, this distinction between public and private is codified in legislation that safeguards private information and the rights of people to keep information private. *Private information* is defined as

> information about behavior that occurs in a context in which an individual can reasonably expect that no observation or recording is taking place and information which has been provided for specific purposes by an individual and the individual can reasonably expect will not be made public (such as a medical record).
>
> (United States Department of Health and Human Services, 2013, p. 9)

Public information includes "any activity – textual, visual, and auditory – (that) is legally available to any internet user without specific permission or authorization from the individual being observed, or from the entity controlling access to the information" (United States Department of Health and Human Services, 2013, p. 9). In many areas of our lives – such as regarding our medical information or our educational records – specific legislation has been established to spell out the rights and responsibilities of users and providers. American constitutional law describes an expectation of privacy. Individuals may expect to have their activities treated as private, mainly when they are in a private space like a home or engaged in private rather than public communications. Privacy allows individuals to restrict others from accessing and controlling information about us.

In the United States, a consensus is emerging regarding the rights that citizens have regarding their privacy and their data privacy. Since the United States is a federal system of government, many laws are made and enforced at the state level. Therefore, not all US states have the same safeguards in place when it comes to consumer and user data (Davis, 2012). However, both Virginia and California have enacted statewide privacy legislation. The law passed in Virginia gives consumers five rights: They have the right to confirm whether a controller is processing their data and to access that data properly. They have the right to correct inaccuracies, have personal data deleted, get a copy of it, and opt out of targeted advertising and profiling ("Virginia Poised to Enact Comprehensive Consumer Privacy Law," February 17, 2021. JDSUPRA.com).

However, some ethicists stress the importance of context in regulating access to people's user data or their data privacy. They argue that you cannot simply adopt blanket rules for how user privacy and data privacy should be safeguarded. They argue that people's comfort levels with data sharing may differ by context. For example, you might not mind having other Peloton users know your name or see your heart rate and other statistics onscreen while you are riding an exercise bike. Still, you might not want this data shared with your employer, nor might you want to share other data with your fellow exercise enthusiasts. They suggest paying attention to contextual integrity, considering the context in which data is shared, the nature of the actors with whom the data is shared, the attributes being shared, and the transmission principles or how the data is shared (Vitak & Zimmer, 2020).

Box 4.1 Going Deeper: Is Privacy Permanent or Temporary?

Privacy is sometimes defined as "the state of being free from observation or disturbance by others." But how long does the state known as privacy endure? Under what circumstances might privacy be violated, and by whom?

This is a question that ethicists are asking about *Direct to Consumer Genetic Testing (DTC GT)*. In many countries across the globe, people can purchase a test kit that they use to swab their cheeks. They then send a DNA sample to a mailing address and receive a link to a site where they can read about what their genetic information reveals about their ethnic and racial origins, and those of people related to them. It can be exciting and fun to learn more about your extended family, including discovering information about new relations, your ethnic and racial profile, and your medical history. But in recent years, it has become clear that your genetic material or DNA doesn't just hold information about you. It may also reveal information about people related to you – people alive now and your ancestors and your descendants (de Groot et al., 2021)

As people learn more about their genetic makeup, they sometimes encounter surprising information about their ancestors. They may find out that their family history about their ethnicity is inaccurate, or they may even find out that they have different parents or grandparents than they thought they did. In some instances, individuals who have placed a child for adoption (or donated genetic material like sperm) with the understanding that this act would be private have found themselves contacted by newly discovered relatives.

Thus far, about 30 million people worldwide have voluntarily provided samples of their genetic material to DTC GT companies (Hazel et al., 2021). As a result of these activities, the following events have occurred:

- In July 2020, law enforcement personnel in the Philippines revealed that they had used DTC GT information to track down men who had come to the Philippines from abroad as sex tourists and fathered children with Filipino women and girls. Law enforcement is pursuing criminal charges against those who had sex with minors and seeking child support from all of these men.
- In 2021, Canadian law enforcement announced that it had used DTC GT materials to solve a 1984 cold case murder.
- In 2020, Baltimore City law enforcement identified a woman who had given birth secretly as a teen and then abandoned her baby, who died of exposure. The crime had occurred ten years previously.
- In a high-profile case, California law enforcement announced that they had used genetic material from a relative obtained from a DTC database to identify the so-called Golden State Killer, a serial rapist who attacked girls and women in the 1970s and 1980s (de Groot et al., 2021).

In each of these cases, an individual was identified because they committed an act that was private at the time but subsequently became public due to DTC GT. These stories indicate how privacy today is increasingly a group activity rather than simply the province of one individual. One's decision to share material in a database can have repercussions in the future and reach back into the past.

Ethicists today have raised concerns about DTC GT for four reasons. First, they believe that individuals might not understand all of the repercussions that might occur (including those in law enforcement) when they give their informed consent to provide genetic material. Second, some ethicists believe that the decision to consent should not belong to one individual alone, since sharing this material may have repercussions for others within one's family. Third, some ethicists believe that if an individual placed a child for adoption many years ago, for example, based on an expectation of privacy surrounding that decision, companies should not assist in violating that historic privacy agreement in the present day. Finally, some ethicists worry about security breaches, pointing out that DTC medical companies do not have the exact requirements and expectations regarding computer security that a hospital, for example, might have.

Is Privacy Permanent?

The notion that privacy can exist at one time but might be withdrawn later (or that privacy is not permanent) is a significant issue in the age of Big Data. As computer engineers become more proficient at building data pipelines that may include many components from many sources, it is increasingly difficult to guarantee that individual data will be kept private. As more and more data is "married up" or aggregated, more unique characteristics of each user will emerge. It is increasingly likely that even anonymized data will be recognized and identifiable someday.

Bibliography

de Groot, N., van Beers, B., & Meynen, G. (2021). Commercial DNA tests and police investigations: A broad bioethical perspective. *Journal of Medical Ethics*, *47*(12), 788–795. Retrieved June 1, 2022, from www.ncbi.nlm.nih.gov/pmc/articles/PMC8639940/.

Hazel, J., Hammack, C., Breisford, K., Malin, B., Beskow, L., & Clayton, E. (2021). Direct-to-consumer genetic testing: Information about ancestry and biological relationships. *PLoS One*. Retrieved June 1, 2022, from https://journals.plos.org/plosone/article?id=10.1371/journal.pone.0260340.

The Challenge: Preserving Privacy While Providing Security

However, citizens and governments face a tremendous ethical challenge in deciding whether and under what conditions to compromise or limit people's right to individual privacy to provide security for a community like a workplace, a city, or a country. Here we recognize that no one has a perfect right to be anonymous. Instead, individuals may be asked to provide information that allows security personnel to engage in *authentication*, defined as "a process that leads us to have a high degree of certainty or probability about the identification of an individual" (Chinchilla, 2012, p. 2). Security personnel often use biometrics – "authentication techniques relying on measurable physiological and individual human characteristics that can be verified using computers" (Chinchilla, 2012, p. 2) – to identify individuals before allowing them access to facilities like a workplace, a government building, or an airplane boarding lounge. Biometric markers are unique to an individual. Computer programs can match an individual's identity by identifying biological information by examining features like fingerprints, retinas, facial geometry, or behavioral characteristics like gait or voice. As Cavoukian et al. (2012, p. 3) point out, biometric data is "unique, permanent, and irrevocable."

Biometric identification technologies allow security personnel to identify who is present at a demonstration or event. Law enforcement officers can track individuals on closed-circuit television surveillance cameras, following them as they go about their daily activities. RFID technologies and GPS tracking technologies in cell phones and other devices create information trails about us as we exist in society. At the same time, security personnel use analytic algorithms to predict people's future behavior based on educated guesses or heuristics related to ethnicity, age, socioeconomic status, gender, and other variables. McFarland (n.d., p. 1) describes how data aggregation technologies may threaten citizen privacy:

> One could rightly or wrongly infer medical and psychological issues from a person's search queries. Sexual activities and preferences, relationships, fantasies, and

economic circumstances. Taken together, they can suggest a comprehensive portrait of a person, including that person's most intimate problems and vulnerabilities.

Cavoukian et al. (2012, p. 3) caution against aggregating databases containing biometric data, arguing that such personal information should not be shared or aggregated without the individual's permission and knowledge.

Profiling techniques can be used for various reasons, including identifying government or private sector employees who might constitute a security risk or even a criminal threat. The American Civil Liberties Union describes criminal *profiling* as "the reliance on a group of characteristics (which are) believed to be associated with crime" by law enforcement officers, who determine who to stop for traffic violations or whom to search based on characteristics such as race (www.aclu.org/other/racial-profiling-definition). These practices contradict Sandel's idea of a person who conducts his life without being prejudged or labeled by others and is responsible for creating his own identity through his actions. Here McFarland (n.d.) writes that such technologies also dehumanize those being evaluated in this way, treating them as collections of attributes rather than as unique individuals. In Chapter 5, on surveillance, we will return to this issue in greater depth.

Box 4.2 Going Deeper: Health Data Privacy

As we think about privacy, we can think about specific types of information that people might like to keep private. In the United States, regulations exist governing education privacy and health privacy. The Health Insurance Portability and Accountability Act of 1996 (HIPAA) allows users to control who may access their health information. Users must give permission before information about their overall health and medical conditions can be provided to an employer, an insurer or other entity. Similarly, the US Family Educational Rights and Privacy Act of 1974 (FERPA) regulates the ways in which educational institutions must safeguard student privacy, and the conditions under which educational information can be shared. Users must give permission before information like a college transcript can be shared with an entity outside of the university, for example.

However, with the advent of the COVID-19 pandemic in 2020, health privacy was one issue that faced new threats. In thinking about the measures a society must take to contain a public health threat, medical officials argued that while medical information privacy might be a private good, public health is a public good. In order to contain a growing threat of viral disease, they thus argued that it might be necessary to ask people to reveal more private health information, at least temporarily (Queen & Harding, 2022).

New Challenges in the Era of COVID

In thinking about challenges to data privacy posed by the COVID pandemic, we can identify two specific instances of interest to those in the field of cybersecurity ethics. First, in the initial waves of the pandemic, officials in both Europe and North America encouraged citizens to download COVID tracker apps onto their mobile phones. Tracking apps made use of **location-based services (LBS)** to provide information about when an individual might come into contact with other infected individuals in crowded public venues. These apps, as they were designed, did not reveal specific information such as the names of infected individuals. Instead, someone who had an app would receive information sent to their cell phone at regular intervals (such as every five minutes). A message from the app consisted of a string of letters and numbers. Everyone in the same location at the same time received the same string.

In some instances, the string was stored only on individual users' phones, while in other instances, it might be stored in a central repository. If someone later reported having COVID, they could inform the app on their phone. The phone would then compare the strings stored on this user's phone with the strings stored on other users' phones. In instances where a user had strings that matched those of the infected person, they would be informed that they had been exposed to COVID. However, no user was informed who specifically might have infected them. In this way, health reporting can occur without compromising user privacy. When information is stored only on users' phones without being sent to a central repository, users are furthermore assured that no one other than themselves is tracking their health. Their health information is thus not being shared with or tracked by the government. Strings are also automatically deleted from everyone's phones after a set period of time.

In addition, many states instituted laws stating that location-based tracker information compiled on a COVID tracker could be used only for the purpose of tracking exposure to this disease. Here, theoretically, law enforcement for example, could use location-based tracking for other purposes – such as determining whether a murder suspect had ever been in the same location as the victim, through comparing the strings on each user's phone. Nonetheless, voluntary compliance with COVID tracker apps was low in many nations – perhaps because users did not understand how the apps worked, and perhaps because users feared that the information might be used for purposes other than disease surveillance, as described here.

A final issue raised by the COVID pandemic is whether users will have changed their preferences in regarding to health data privacy as a result of having lived through the pandemic. That is, some analysts suggest that users might experience a change in how they view their health data – and may be permanently more willing to share private health data for the good of public health. Such cultural changes may occur unevenly across the globe, with some nations changing the ways that they now think about health data privacy, while others may stick to the old ways of thinking about health data as a type of private property.

Bibliography

Queen, D., & Harding, K. (2022). Data capture, analysis, utility and privacy and a covid legacy. *International Wound Journal*, *19*(3), 465–466.

Simpson, A. (2022, February 26). Scottish government breached data privacy laws with covid passport app. *The Herald*, p. 6.

Privacy and Justice

Van den Hoven (1997, p. 31) argues that privacy is essential for people to experience equity and justice. Everyone should have an equal ability to decide what to keep secret and what to reveal, regardless of factors like socioeconomic class. Floridi describes privacy in terms of ownership. If we own our information about ourselves, we can decide whether and under what circumstances to give that information away. In an ethical world, everyone would have an equal ability to own their personal information (Tavani, 2008).

However, surveillance or spying often violates this principle of equality in which everyone has a right to keep their secrets private. Some people are more likely to be monitored, even if they rob them of dignity and independence. Most people agree that young children should have less privacy since they need to be watched and monitored closely because they lack adult judgment. Those elderly or infirm may also have less privacy as they are taken care of in facilities, often by paid caretakers. In these instances, those in authority have made a judgment call if there are trade-offs between privacy and the provision of security.

New technological developments also have ramifications for privacy and equity. We may decide to trade away some of our privacy (for example, information about purchasing decisions) in return for a discount on groceries or some other product. Others may decide to spend more money to, for example, hire an online reputation management company to manage more carefully what information strangers can find out about them through online searching (Van den Hoven, 1997, p. 31). This has been referred to as a *privacy calculus*. In thinking about these developments, we need to be vigilant against developing a system of *tiered privacy* in which some individuals are monitored or surveilled more closely than others.

Privacy and Trust

Privacy also helps to create trust. Individuals are more likely to invest with banks or to engage in relationships with institutions where they are assured that their information will be kept confidential, that it will not be compromised through a data breach, and that others will not have access, including the ability to copy and disseminate that information. In this way, Brey (2007) notes that privacy is also related to property rights, describing actions like copyright theft and information piracy as violations of user privacy.

Thus, as we have seen, privacy is a foundational value because so many other rights and values depend upon the provision of privacy. Figure 4.1 depicts the cluster of related activities and values associated with privacy.

It is challenging for governments and corporations to create and enforce regimes or sets of rules regarding what personal data about people will be captured and stored, who may access it, and under what circumstances, given the prevalence of technologies like ubiquitous computing. **Ubiquitous computing** refers to the practice of embedding technology in everyday items. They may collect and transmit information to other objects, often within the user's awareness. The Internet of Things relies on ubiquitous computing to establish an environment where the objects in your home can "talk" to each other, with your thermostat, for example, talking to your smartphone.

Marmor (2015, p. 3) notes that to secure people's privacy, it is necessary to create a "reasonably secure and predictable environment" in which individuals and groups

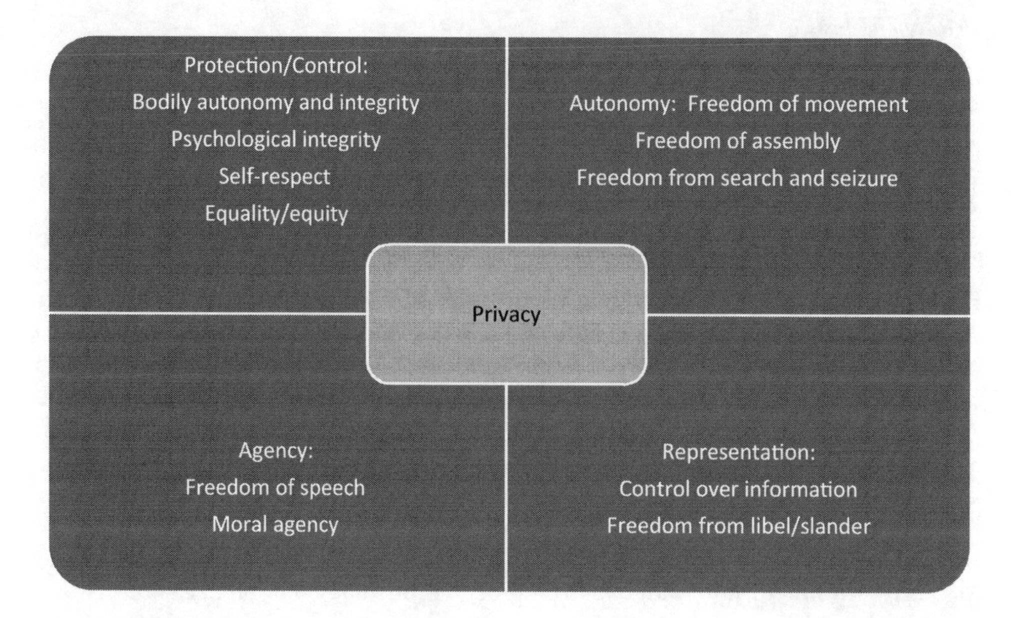

Figure 4.1 Rights and Values Associated With Privacy

understand how information is organized, how it flows, and the consequences which might result from the release of information. Thus, individuals involved in information technology and cybersecurity play a vital role in safeguarding privacy by helping build, establish, regulate, and secure environments in which information flows.

Best Practices for Privacy Preservation

How, then, can those who work in cybersecurity help to safeguard user privacy? The privacy by design paradigm for software design emphasizes that it is far better for practitioners to think about how best to provide user privacy from the beginning rather than to seek to "graft on" privacy-preserving technologies once the product has been created. *Operational definitions of privacy* refer to the specific capabilities within technology for safeguarding data. An operational definition of privacy might thus specify what privacy settings are available in a program and who can administer these settings.

The European Union's General Data Protection Regulation incorporates data protection principles by design and default. Data Protection by Design means those who create new data architectures, apps, or data storage facilities should prioritize and consider data protection from the earliest stages as they create their products. The onus, the stress, is not on consumers to protect their data or governments. Instead, each corporation in the EU must appoint a data controller whose responsibility is to ensure that the corporation is prioritizing and responding to the privacy needs of the data subjects using its products (van Dijk et al., 2018).

To preserve user privacy, developers, in particular, should consider privacy questions as they create their data architecture. Hoepman (2013) suggests that developers should consider the following principles as they do so: First, they should make sure that they are working with only the minimal amount of data necessary to carry out a particular operation. Suppose they do not require access to some portion of a user's data. In that case, they should strive to partition or classify that data rather than make it all available for a particular operation. Second, developers should hide personal data and their relationships, making it accessible only to those who require it. They should also process personal data at the highest level of aggregation with the least possible detail required.

Whenever possible, users should be informed that their data is being collected and processed, and they should be asked to consent. (Here, specific legislation such as the European Union's General Data Protection Regulation may require that specific consent processes be followed.) Data subjects should thus have control – consenting to sharing some types of information and refusing consent for others (Hoepman, 2013).

Privacy: Can It Evolve and Change?

Some analysts describe privacy as a universal value that all cultures recognize (Bok, 1989). However, others argue that privacy is a *social construct* that varies widely across time and cultures (Boyd & Marwick, 2011; Tuffley & Antonio, 2016). In this view, individual and group ideas about privacy might change over time in response to social changes or even technological changes like the widespread availability of social media. Danah Boyd has looked at how teenagers think about privacy due to social media exposure, arguing that young people may no longer recognize the previously accepted clear distinction between the homes as a private space demarcated from more public spaces.

We use normative expectations of privacy to refer to ideas about what is socially appropriate in terms of privacy, though normative expectations of privacy often line up with legal understandings. Because norms – or unwritten social expectations or rules – are often specific to a culture, norms regarding privacy may also differ from one society to another (Belloti, 1997; Etzioni, 2015). For example, in many Middle Eastern cultures, women may wear a hijab to control who may view their bodies. In other cultures, norms regarding viewing the female body may differ, and clothing choices will reflect this difference.

Analysts disagree about data providers' ethical responsibilities to guarantee data privacy, given technological limitations. Should a data provider be required to guarantee citizens that their data will be completely safe, and is this even possible? Kligiene (2012) notes that currently, there is no universal solution for guaranteeing data privacy to citizens in all forms of social media. Aicardi et al. (2016) suggest that just because there is no technique for tracing all of the content you create online back to you directly, this does not mean that there will not be a technique for doing so in the future. They suggest that the most a data provider can say is that "at present, your data is safe from intrusion and your anonymity is secure – but that does not mean that it will continue to be into the future."

Is There Such a Thing as Too Much Privacy?

While, thus far, we have made the case that privacy is healthy, socially desirable, and necessary, not everyone agrees with this view. Plato, whom we met in Chapter 1, saw privacy as an extreme counterproductive value in an ideal state. In *The Laws*, he wrote, "Friends have all things in common" (1967, Vol. 1, 5). Famously, US president Woodrow Wilson argued that people who request privacy or secrecy often hide something, drawing upon the thoughts of the English philosopher Jeremy Bentham. Wilson equated secrecy with impropriety, particularly in the case of governments seeking to keep secrets from citizens.

He thus prized transparency or openness against secrecy as a value, running a 1912 presidential campaign in which he spoke about openness, sunshine, and fresh air, equating secrecy with corruption (Bok, 1989).

Legal rulings have established that privacy is the right individuals have. However, corporations and government agencies cannot use privacy rights to disregard or ignore their obligation to share information about programs and policies with constituents. Here, the cooperative value or norm is one of **transparency** – or the obligation to share information with citizens. In the United States, citizens can file Freedom of Information Act (FOIA) requests to get permission to see government documents (Ballotpedia). Prospective voters may also feel entitled to information about a candidate's health or finances. Local, state, and federal governments often participate in *Open Data Initiatives*, moving their public records online and making them searchable so that citizens can see their public officials' salaries and how their tax dollars are being spent by examining budget documents.

While governments and public officials can expect less privacy, even individuals acknowledge that the right to privacy is not absolute. We might still be compelled to furnish information about our finances for functions like paying taxes, writing a will, or negotiating a divorce. We also do not have an absolute right to communicate anonymously, engage in untraceable activities, maintain sites, or carry out activities on the so-called Dark Web.

The fields of *digital forensics* and *encryption* rest on this assumption that privacy is not absolute and that we do not have an absolute right to keep secrets – since not all secrets are good secrets. Individuals with expertise in digital forensics assist law enforcement with retrieving data from captured digital devices. They might attempt to access data on a captured computer regarding drug transactions, look for evidence that an individual was engaged in the pornography trade, or look for evidence of illegal financial transactions such as embezzlement or money laundering. Digital forensics experts might testify in court regarding their findings, and several companies and schools now offer certification in this growing field.

Other computer experts may work in the field of encryption, devising robust encryption and decrypting data, assuring the safety of individuals' and corporations' legitimate financial transactions, thus enabling the US and international economies to function routinely and without interruptions.

Balancing Competing Needs for Personal and Data Privacy

Earlier in this chapter, we introduced some best practices that developers and data architects can follow to preserve user privacy. But often, computer specialists will face the problem of competing priorities, as well as the problem of competing audiences.

In thinking about whether people have the right to health information privacy during a pandemic like COVID, it becomes clear that a doctor has multiple "customers" or clients that she must consider as she decides whether or not to share health information data. While a patient might prefer that all of their health information be kept confidential, the doctor may find that she also has an ethical or moral obligation to the larger community to safeguard public health – which might involve releasing some limited amount of individual data if, for example, her patient has an infectious disease. The doctor must consider her obligations as a member of a specialty or profession, including her obligations to her patients and the larger community. In such situations, it may be challenging to sort out a values hierarchy or determine which obligations are primary.

In the same way, those who work with information and data also have multiple audiences to consider as they make decisions about user privacy. The data architect or engineer may have specific obligations to the data subject (to preserve their privacy), their employer (to produce an efficient product that works well for a reasonable cost), and their community. The quest for privacy and utility are thus often viewed as zero-sum. To better preserve user privacy, safeguards might be put in place, which allow for less efficient data storage or data usage, while to use data most efficiently, user privacy might be compromised (Wu et al., 2021).

In addition, some analysts worry that accuracy may suffer in prioritizing privacy as a value in creating a software program or package. In particular, some analysts argue that the best way for a machine learning algorithm to learn how to better engage in practices like facial recognition is to run, even if it might make incorrect identifications in the early stages. They note that an AI can't fix itself if it doesn't know it makes errors (Amoore, 2021). Thus, some utilitarian ethicists might describe risks to user privacy as a necessary cost to create a better product in the long run.

Figure 4.2 illustrates the trade-offs between privacy and other ethical values, which developers, data architects, and engineers must navigate.

We can identify several regulations that safeguard individual privacy, including privacy regarding data provision. Regulations may be created by local, state, and national bodies and occupational bodies like the American Institute of Certified Public Accountants. In addition, the term *fair information practices* refers to principles adopted globally to ensure that individuals' privacy rights are protected and that corporations take sufficient care to ensure that they do so (Culnan & Williams, 2009).

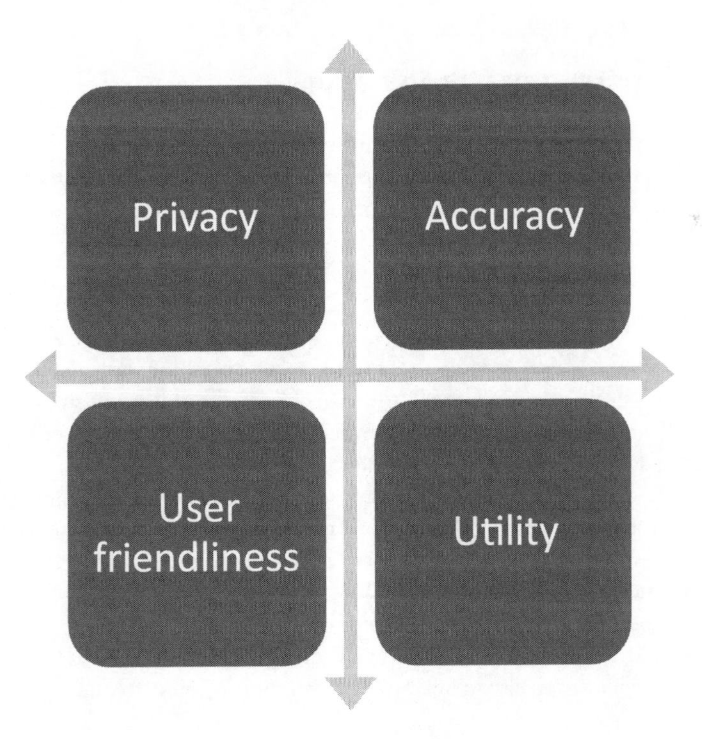

Figure 4.2 Protecting Citizen Privacy

In American law, violating someone's privacy constitutes a type of harm and is thus the subject of tort law, a branch of civil law that deals with the actions of providing damages for a wrong. In the 1965 Supreme Court case *Griswold v. Connecticut*, Justice Douglas stated that the legal right to privacy is implied in four amendments to the US Constitution: the First Amendment (providing freedom of speech and the press), the Third Amendment (rights establishing the sanctity of one's home), the Fourth Amendment (rights against unlawful search and seizure) and the Fifth Amendment (right against having to incriminate oneself).

Many types of information are regarded as sensitive or private. Specific legislation lays out what information may be collected and stored, who may have access to that information, and under what conditions.

One area where privacy rights have become controversial in recent years is that of hiring situations. In the United States, labor relations laws are stringent regarding what questions are off limits in a job interview. Employers cannot ask about prior arrests, plans to get married or get pregnant, or religious observance (Giang, 2013). However, employers can now thoroughly investigate applicants' private lives using social media in order to determine their age, race, gender, political views, photographs, and qualifications (careerbuilder.com). Legal cases have supported the finding that a prospective employee has a lower expectation of privacy from an employer than an employee who has a relationship with the employer (Sloan et al., 2012, p. 6). In addition, employers have a fair amount of discretion regarding monitoring employees' social media use in the workplace once hired. For this reason, experts advise job candidates to sanitize their online profiles and suggest that even once hired, employees should be careful about posting provocative photos, making off-color remarks, or sharing information about their employers.

Box 4.3 Critical Issues: Brain Hacking

Today some ethicists are concerned about threats to "brain privacy." They describe technological developments that may someday allow others to read our brains' contents and know what we are thinking and even dreaming about without our awareness or consent.

How Does It Work?

Until recently, most users viewed technology as something outside their physical bodies. However, it is now possible to begin erasing the distinction between your physical self and technology. Today, some technology (such as medical devices) might be embedded or implanted within an individual.

One common type of embedded technology is a *neural* or brain interface. Today some medical patients work together with technology to overcome physical limitations. A transmitter might be placed in someone's body to read electrical impulses sent by their nerves. A computer program can then read these impulses or messages, enabling the user to communicate a thought. People may also use a neural interface to control their muscles to walk or move their arms. At the same time, deep brain stimulation devices may also determine if someone is about to have a seizure and then send a signal to the brain to disrupt the seizure. In the future, neural interfaces might be used not merely as treatments but also as a neural enhancement, enabling people to communicate faster, or to write or type more efficiently.

User data can be collected and analyzed using both embedded and wearable technologies. In addition, technology users might use an external device – like a Magnetic Resonance Imaging (MRI) machine or an electroencephalogram (or EEG) to measure individuals' responses to inputs such as images displayed on an external screen. This use of technology has been referred to as *brain fingerprinting* or *brain-reading*. Using an external device that a user must submit to, it is theoretically possible to "see inside someone's brain" or measure their emotional and cognitive responses in various scenarios. Scientists are also exploring technologies to map your brain's responses while you sleep to be someday able to watch your dreams (Mecacci & Haselager, 2019).

Using devices and technologies like embedded devices, wearable devices, and brain fingerprinting, then, it is possible that in the future, users might not fully experience the privacy of owning what goes on in their minds – since even this information could one day be shared, compiled, displayed, and analyzed by others. Such technologies present new ethical and legal challenges.

Privacy Dilemmas: Legal Issues

One group of individuals interested in the science of brain-reading is criminologists (Alpert, 2007). As Kerr et al. explain, new neuroimaging technologies will eventually allow law enforcement to dispense with traditional interrogations and even with so-called lie detector tests. Instead, law enforcement officials might show you a photo of a crime scene and monitor your brain to see if it "recognizes" the image (since old images are stored in one part of the brain, while a different part of the brain reacts to a brand new image). If you recognize the image, it might prove that you had visited the crime scene before (Kerr et al., 2008).

Ethicists are also worried about how brain imaging can make inferences about someone's intent. For example, in a security clearance interview for a government job, you might be asked, "Are you seeking this job to harm the United States?" Here, a technician could monitor whether the part of your brain which makes up lies was activated. In addition, scientists note that one's brain works harder when telling a lie than it does when telling the truth, producing more energy – which can be measured.

A Brain-Computer Interface (a device that acquires and translates neural data, applying it for the control of other systems) might also be used by law enforcement to gather data not just on an individual's activities but also on an individual's preferences. For example, law enforcement could tell if a suspect was sexually attracted to children by displaying images to the suspect and tracking how their brain responded to the images being displayed. It would be possible to tell if someone harbored racist or misogynistic attitudes or even identify specific sexual identities – even if the subject never chose to act upon those attitudes. But should a future employer or a law enforcement agency be able to preclude you from job considerations, or suspect you of a crime, based on preferences and attitudes rather than actual activities? (Wahlstrom et al., 2017).

Other professionals, however, argue that brain surveillance is more objective than any other method of measuring whether someone is a security risk. Perhaps it is better to have an algorithm or program decide if you are a security risk rather than relying on a faulty human to make that determination.

Here, some analysts believe that the US Health Insurance Portability and Accountability Act (HIPAA) means that information obtained by brain reading – such as an MRI or CAT scan – constitutes legally protected individual health information. Therefore, an individual could not be compelled or required to furnish this information to any outside researcher or health professional without their consent.

One might also argue that searching someone else's brain violates the US constitutional right to not be the subject of search and seizure without a warrant. In Canada, researchers have suggested that spying on someone's brain might violate the Charter of Rights and Freedoms, Section 8, which provides the right to be secure against unreasonable search and seizure. Other Canadian rights activists have to reference Section 11c of the Charter of Rights and Freedom, which provides citizens with a right not to engage in self-incrimination. Section 7 provides a right to the security of one's person.

Other scholars argue that such technology could violate the Fifth Amendment to the US Constitution, which allows a defendant to receive due process. Criminal defendants have a constitutional right to remain silent and refrain from self-incrimination. If an interrogator "reads" your brain and sees that you have prior knowledge of a crime, would doing so violate your right to remain silent and refuse to testify against yourself? (Mecacci & Haselager, 2019).

Privacy Dilemmas: Ethical Issues

Ethicists warn of a future where we may not fully control our thoughts and minds. They argue that this may not always be true because today, one must give consent for someone else to enter our brains (Bonaci et al., 2015). However, they describe a future where it might be possible to engage in "remote surreptitious brain surveillance." It might be possible to merely monitor another person's brain activity by being in their presence. You would not need to be in a hospital laboratory to be monitored, you would not need to consent, and you might even be monitored without your knowledge (Kerr et al., 2008).

And once someone has entered our minds and thoughts, what might they do there? Some futurists warn against emerging "neuro crimes" in which individuals might find that their thoughts and brains have been hacked without their knowledge or consent (Mecacci & Haselager, 2019). As a result, people might engage in activities like purchasing products due to stimuli they receive without their consent or awareness. Imagine a future in which a hacker can change people's preferences – perhaps causing them to support political causes or candidates without their knowledge or consent.

Heslen (2020) suggests that in the future, neuro-hacking might be used in military psychological operations to, for example, compel a group of people to surrender to their occupiers through "manufacturing consent." But does one individual have the right to "steer" other people's brains and decision-making in this way? Does doing so constitute "playing God" in some way?

As noted earlier in this text, ideally, those who design devices like neuro-cognitive interfaces have an ethical responsibility to consider such dilemmas before consenting to create a technology. *Brain-Computer Interfaces*, for example, could be designed to be privacy-preserving (Wahlstrom et al., 2017). We may expect engineers to come together to define basic ethical principles regarding consent, surveillance, and awareness of when these technologies are being applied and implemented, and the conditions under which they can and cannot be used appropriately.

Bibliography

Alpert, S. (2007). Brain privacy: How can we protect it? *American Journal of Bioethics*, 7(9), 70–73.

Bonaci, T., Calo, R., & Chizeck, H. (2015, June). App stores for the brain: Privacy and security in brain-computer interfaces. *IEEE Technology and Societ Magazine*, *34*(2), 32–39.

Heslen, J. J. (2020). Neurocognitive hacking: A new capability in cyber conflict? *Politics and the Life Sciences*, *39*(1), 87–100. Retrieved June 1, 2022, from https://pubmed.ncbi.nlm.nih.gov/32697058/.

Ienca, M. (2015). Neuroprivacy, neurosecurity and brain-hacking: Emerging issues in neural engineering. *Bioethical Forum*, *8*(2), 51–53. Retrieved June 1, 2022, from www.bioethica-forum.ch/docs/15_2/05_Ienca_BF8_2.pdf.

Kerr, I., Binnie, M., & Aoki, C. (2008). Tessling on my brain: The future of brain privacy in the criminal justice system. *Canadian Journal of Criminology and Criminal Justice*, *50*(3), 367–387. Retrieved June 1, 2022, from https://doi.org/10.3138/cjccj.50.3.367.

Mecacci, G., & Haselager, P. (2019). Identifying criteria for the evaluation of the implications of brain reading for mental privacy. *Science and Engineering Ethics*, *25*(1), 443–461.

The Committee on Science and Law. (n.d.). *Are your thoughts your own? Neuroprivacy and the legal implications of brain imaging*. New York: New York City Bar Association.

Wahlstrom, K., Fairwether, B., & Ashman, H. (2017). Privacy and brain-computer interfaces: Method and interim findings. *The Orbit Journal*, *1*(2), 1–19.

Preserving Data Privacy

Thinking about the ethics of privacy in a virtual environment is complicated, however, not just when we ask about identities, including our digital identities, but when we begin to ask about our data. The term *digital footprint* refers to the data we use and produce on all of our devices when we are online. As Kligiene explains:

> A digital footprint is a combination of activities and behavior when the entity under consideration (a person or something else) acts in the digital environment. These may

> be log on or off records, address of visited web pages, open or developed files, e-mail or chat records.
>
> (Kligiene, 2012, p. 69)

In his work on information ethics, Floridi (2013) suggests that we own our digital footprints and that they are our digital footprints authentically. He suggests we feel harmed if our data is taken, shared, or stored without permission. We will feel equally violated by the thought of someone peeking at our data as we would if they physically peeked into our homes without our knowledge or consent. We would also feel equally violated if someone stole our virtual identity and proceeded to act in cyberspace as though they were us (Tavani, 2008).

But providing online and data privacy is challenging for three reasons: First, connectivity and networking have eroded the distinction between public and private space and public and private activities. In addition, our ability to store, share, and aggregate data creates new privacy problems. Finally, developments in biometrics are eroding the distinction between our physical bodies and our digital bodies and identities, and creating new ethical challenges.

Problem 1: Connectivity and Networking

Our information environment is characterized by *connectivity*, or the ability to make connections between points in a telecommunications system. We might furnish data to one particular website, but it might be shared with other entities within a network, often without our knowledge or permission. And the distinction presented in the earlier section "Private Space and Public Space" does not map neatly onto cyberspace. Instead, activities are characterized by a lack of clear boundaries between public and private spaces and public and private activities.

This distinction is particularly unclear in the area of social media. Sloan and Warner (2014) define *social media* as "any electronic medium where users may create, share or view user-generated content, including videos, photographs, speech or sound" (p. 1). But does this social content you generate – meant to be shared and viewed by others – belong to you alone, or does it belong to other interested parties, like your employer or law enforcement agencies? Can we speak of a right to privacy on social media?

This erosion of the public and private distinction poses several ethical and legal issues: Under what circumstances do we have a right to privacy online? Are there situations where we should be required to relinquish this right to privacy, and under what conditions? On the one hand, we may believe that as individuals, we should have the right to say and do what we want online, no matter where we are or what we are doing. But on the other hand, we also expect to feel and be safe in our workplaces and communities, to be protected from harassment and bullying, and for authorities to be aware of dangers and threats to our safety and perhaps even to act preemptively to protect us and keep us safe.

Problem 2: Privacy Versus Security

Authorities – including employees – have the responsibility to police the safety of their environments, even if doing so means accepting that privacy violations will occur. Individuals and groups may need to monitor individual social media activities to safeguard individuals in the real world. In this view, the risks associated with allowing a disgruntled employee or even a terrorist to threaten individuals or carry out dangerous activities are so significant that it is worth violating individual online privacy to secure the safety of those affected by an individual's online activities.

US legal developments support this ethical understanding that employers have a right and responsibility to monitor employees' personal social media posts – reacting to defamatory posts and those that show danger or disclose private or proprietary information. In the United States, the Electronic Communications Privacy Act establishes the principle that we do not have an absolute right to privacy online during working hours, whether we are physically or virtually located at a workplace. This law allows employers to monitor how their employees spend their time online during the working hour – tracking the websites visited and the activities carried out. However, new laws are beginning to limit these rights. In Maryland, California, and Tennessee, employers cannot ask employees for their names and passwords for social media accounts nor can they demand that they be added to accounts (Sloan & Warner, 2014, p. 23). In Tennessee, employers are allowed to conduct activities that assure that they treat company data as confidential, including searching for postings in their name on social media. However, the legislation specifies which employer activities would constitute an invasion of privacy and encroachment on citizens' First Amendment rights to free speech. In addition, some employee unions have written privacy clauses into their contracts in Tennessee and other states, specifying what employers may and may not ask for or monitor (workplacefairness.org).

The Duty of Individual Users

In addition, many analysts emphasize the principle that internet users have a responsibility to proactively take steps to safeguard their privacy and the integrity of their data online. They cannot simply expect that their internet service provider or the website owner will take the necessary steps to safeguard their data and thus need to be individually motivated.

Floridi (2013) suggests that all information technology users should have a shared commitment to protecting the "information infosphere." His argument shares common ground with environmentalist principles in arguing that information environment users (like physical environment users) should seek to preserve an environment and consider the effects their actions might have on that environment. Information environment users thus might refrain from engaging in practices like spamming, harassment, and trolling. It may also mean taking a serious interest in practicing good cyber hygiene, from being aware of phishing practices, investing in a good firewall, and investigating and using privacy-enhancing technologies (PET) like authentication, biometrics, and encryption.

Problem 3: Data Storage and Sharing

However, while individuals may be careful about privacy in their online interactions, they often don't consider how their data is stored or shared. **Ubiquitous computing** refers to the practice of embedding technology within everyday objects so that they can store and collect data, sharing it with other objects within the Internet of Things to which they are connected. The technology in these objects is hidden, and users are often unaware that this data is being collected, stored, and shared (Belloti, 1997).

Data mining is searching through databases to discover relationships and patterns. As Al-Saggaf and Islam (2015) note, "data mining can be used to classify and categorize, and to know more about a user and to make predictions about his future behavior –

such as future purchasing behavior" (p. 946). *Web usage mining* looks at users' behavior, including what websites they visit and look for online. These practices mean that if you visit the website for a university, the university's experts can tell what other pages you looked at first and perhaps identify other universities you might be considering for your program.

Thus, we must consider the ethical obligations of those tasked with collecting, storing, or managing data. What standards should they adhere to, and what should be the penalties for not doing so? What should be their guiding principles and values, and who should they regard as their ethical subject? To whom are they answerable?

The essential condition to which those entities collecting data must adhere is the notion of **informed consent**. This idea is borrowed from the standards and protocols established within the scientific community for so-called *Human Subjects Protocols*. These conditions are laid out in the Belmont Report, a federal document issued in 1978 by the National Commission for the Protection of Human Subjects of Biomedical and Behavioral Research. This report spells out the rights of individuals regarding biomedical and behavioral research – and was updated in 2010 to include internet research. The document uses a virtue ethics approach to argue that all research should be conducted with an eye toward respect for people (including their autonomy, courtesy, respect), beneficence, and justice (United States Department of Health and Human Services, 2013; Sims, 2010). Researchers must consider the dignity and feelings of human subjects in designing an experiment. At universities, researchers typically present a petition to a *Human Subjects Review Board* in which they spell out how they will meet specific conditions. Conditions may be federally mandated, state or locally mandated, or mandated by the university itself.

Such conditions usually include the requirement that subjects are informed about the experiment's purposes, what they can expect to encounter in the experiment (including any side effects); how their anonymity will be maintained when researchers report the experiment's result, and how data collected will be stored, maintained, or destroyed. They are also told of their rights within the experiment – such as the right to stop participating. Then they must give their written consent to participate in the study. The consent notice might spell out conditions such as whether data provided by users could be sold or shared with other corporations or websites. It might also include a statement regarding acceptable and unacceptable behaviors online and the conditions under which a user might be criticized or depermitted from a site.

Human Subjects Review protocols and their internet equivalent, the consent statement, help preserve the subject's rights. But several problems presently exist with consent statements: First, many are long and unwieldy, and subjects may not read all the dense languages. They might sign their rights away without fully considering what they agree to (Duvall et al., 2016).

Second, those with limited English proficiency, low literacy, or cognitive impairments, including the elderly, might struggle to understand user agreements. Many people frequently do not understand what they are signing (Etzioni, 2015, p. 1269). Finally, some issues addressed in consent agreements might be novel problems that have not been fully addressed since they have not happened. For example, in the future, we may encounter scenarios where a tech company goes bankrupt, and in settling its accounts, its assets might be sold off. If their primary asset is data that has been collected about users, this data might be sold to a third party.

Some analysts feel that because of the seriousness of these problems, those who collect data often commit ethical breaches in deceiving or coercing individuals to provide data.

They note that citizens cannot opt out of most consent agreements since they may require a cell phone to do their job. Culnan and Williams (2009, p. 681) suggest a fundamental power imbalance between the citizens who furnish data and the corporations that collect it. Citizens know less about technology, issues, and rights than corporations might, and thus, privacy harms are due to this power imbalance.

And Ellerbrok (2011) describes how users might decide that new technology is harmless or even a fun "game" due to the circumstances under which it is introduced. For example, she suggests that facial recognition (FR) technology has military applications, such as scanning a crowd to identify individuals on the United States' Terror Watch List. However, because most people encounter FR when it tags their friends on Facebook, they do not associate it with military applications. Here we might call the technology's creators deceptive, since they created acceptance of the new technology by introducing it in this way without making sure that users understand its applications.

Box 4.4 Tech Talk: Techniques and Mechanisms for Insuring User Privacy

Technology users often engage in behaviors that collect data about their movements, purchases, and browsing history. In many instances, they might not be aware of the programs that run on their wearable devices or cell phones and may not think very much about how these programs are configured. But programmers, developers, and user interface (UI) and user experience (UX) designers think extensively about these issues. Every day, they decide the types of information devices collected, how they are stored, and the choices users might encounter in configuring their devices. This short article presents some of the issues designers encounter in safeguarding user privacy and user data privacy.

Creating Privacy Risks

Imagine you visit a location while carrying your cell phone. Your phone preserves spatial data about your travels. Spatial data is any data extracted into information identifying a particular place, using coordinates like longitude, latitude, districts, or streets (Calleo, 2021). Or perhaps you upload an image to social media or tweet while in a location with your location-based data services enabled.

In each case, you have created a privacy risk. *Open privacy risk* is a situation in which someone can access your data. In contrast, *deductive privacy risk* is a situation in which someone can learn private information about you by combining multiple datasets.

Presumably, someone who breached your data would now know where you were physically located in space at that time. Or they could find out such information ethically and legally – through interfacing with a service like Twitter which provided this georeferenced information through an API.

But now imagine that there was only one institution within these coordinates – a cancer hospital, a mental institution, or a military base. What information might an adversary now have about you and your habits?

Location Privacy Preserving Mechanisms (LPPMs)

App developers aware of this problem have developed a "semantic cloaking fix." Using *semantic cloaking regions* (*SCRs*), developers can ensure that multiple different kinds of entities are located within a region when location-based data is preserved. (They draw a circle around your location, widening it until it includes multiple different types of sites. They might also draw a more extensive region to ensure that it contains multiple users so that one specific user's movements cannot be tracked.) Semantic cloaking can help preserve user privacy, though it is not foolproof. When you regularly visit a location, it may still be possible for an attacker to home in on exactly where you go by looking at your data over time (Riaz et al., 2017).

Other location privacy preservation mechanisms (LPPMs) may include the creation of "dummy locations" randomly generated by a system or using **location perturbation mechanisms** such as reporting on a noisy location with multiple users rather than the actual location. **Differential privacy**-preserving mechanisms systematically alter the data stored by performing mathematical calculations. The technology is similar to message encryption in that a key can be used to decrypt the data later if necessary (Schneier, 2017).

Pros and Cons of LPPMs

There are many positive reasons for using, applying, and developing LPPMs. In autocratic regimes where people might be punished for attending a political demonstration, for example, LPPMs can help assure people's safety and their valuable right to engage in free assembly.

However, altering data can create other types of problems for other users. For example, applying LPPMs may make user data less useful for researchers in other areas – like urban planners who may use location data to decide where to site a hospital or whether or not to build more roads. This is particularly true in situations where a planner, for example, might wish to make an analysis based on aggregating multiple datasets about user movements and locations. It would not be easy to aggregate datasets, all of which have had different calculations performed to alter or modify data to preserve user privacy. In addition, when law enforcement might wish to examine a user's data to use it as evidence gathering, LPPMs can also render the data less valuable and less exact.

In this way, developers must consider the trade-offs between the privacy granted by using protocols like SCRs. Therefore, the data collected is less valid or efficient for other users (Li et al., 2019).

Future Advances

The European General Data Protection Regulation mandates that European users consent (by ticking a box) before collecting their location data. Therefore, in Europe, there is an incentive to apply methods that preserve user privacy. In contrast, within the United States, individual states have regulations regarding what sorts of location data can be collected and stored and who may have access to that data (such as law enforcement). If privacy-preserving mechanisms are being applied in the United States, they are applied in a less organized and more haphazard manner.

Bibliography

Calleo, Y. (2021). Geo referenced social media data and privacy: The case of Twitter. *Academia Letters*, Article 2679. http://doi.org/10.20935/AI2679.

Li, B., Zhu, H., & Xei, M. (2021). Quantifying location privacy risks under heterogeneous correlations. *IEEE Access*. Retrieved June 1, 2022, from https://ieeexplore.ieee.org/document/9343824.

Li, Y., Cao, X., Yuan, Y., & Wang, G. (2019). PrivSem: Protection location privacy using semantic and differential privacy. *World Wide Web*, *22*(1), 2407–2436.

Riaz, Z., Durr, F., & Rothermel, K. (2017, November 7). Understanding vulnerabilities of location privacy mechanisms against mobility protection attacks. In *14th EAI international conference on mobile and ubiquitous systems: Computing, networking and services*. Stuttgart: Association for Computing Machinery. Retrieved June 1, 2022, from https://eudl.eu/doi/10.4108/eai.7-11-2017.2273757.

Schneier, B. (2017, July 18). Surveillance is the busness model of the internet. *Schneier on Security*. Retrieved June 1, 2022, from https://doi.org/10.1186/s40537-020-00318-5.

However, others emphasize that users are also responsible for informing themselves and understanding and using privacy-enhancing technologies to safeguard their data. Burkert defines **privacy-enhancing technologies (PETs)** as "technical and organizational concepts that protect personal identity." PETS can include encryption technologies and measures like digital signatures and pseudonyms." PETS can thus allow users to have more control over how much of their information is revealed and in what format.

An Ethical Approach to Data Storage

What, then, might an ethical approach to data storage look like? Here we assume that data is ethically neutral – neither ethical nor unethical. What is ethical or unethical is the decisions managers make regarding that data – what to collect, how to store it, and how to treat stored data. In this regard, Etzioni suggests that we consider three factors – the volume of information collected, the level of sensitivity of that information, and the degree of "cybernation" taking place.

Here the term *cybernation*, based on wordplay from the term *hibernation*, refers to the storage or stockpiling of information, including compiling a dossier and sharing information between agencies and groups (Etzioni, 2015, p. 1273). Etzioni notes that there is a difference between, for example, collecting a small volume of information about someone (such as their answers to a survey) and engaging in the collection of bulk data. Here, *bulk data* is defined as "information from multiple records, whose primary relationship to each other is their shared origin from single or multiple databases" (Information Resource of Maine, n.d.). In addition, Etzioni distinguishes bulk data collection from that of *metadata* – defined as "Structured information that describes, explains, locates or otherwise makes it easier to retrieve, use or manage an information resource" (NISO, 2004, p. 1). Metadata is specific data related to a user, arguing that reading all of someone's e-mail correspondence, for example, is more intrusive than merely running an algorithm that looks for keywords in an extensive collection of data.

In addition, Etzioni suggests that collectors have a particular responsibility to safeguard sensitive information that might be of an intimate or highly intrusive nature:

> Not all personal information can or should be accorded the same level of protection. The more sensitive the information an agent seeks to collect, the more measures to protect privacy should be implemented. The higher the public interest must be before collecting the information is legitimate.
>
> (Etzioni, 2015, p. 1278)

In addition, those who collect data have an ethical obligation to inform citizens in situations where a privacy breach may have occurred. A privacy breach is defined as "the loss of unauthorized access to, or disclosure of, personal information" (Office of the Privacy Commissioner of Canada). In the United States currently, this obligation is spelled out in state-level legislation, with Congress currently debating laws to establish a national standard for data breach notification. The Directive on Privacy and Electronic Communications (E-Privacy Directive) includes a breach notification law in Europe.

Finally, corporations have an ethical obligation to carry out privacy **impact assessments** (PIAs) before establishing new systems or uses of personal information. A PIA may examine whether proposed changes in a system comply with legal requirements and examine the risks associated with collecting and using personal information. It may also propose mitigation measures in situations where a breach occurs (Culnan & Williams, 2009). The E-Government Act of 2002 requires that US federal agencies carry out a PIA before developing public information systems.

The Council for Big Data, Ethics, and Society is currently proposing measures to preserve subjects' data privacy. Such measures require a "*data ethics plan*" in grant applications such as those administered by the US National Science Foundation (Metcalf, 2014).

Cybersecurity Data Science: Learning from Privacy Breaches

Cybersecurity experts can increasingly learn from reported data privacy breaches as data privacy develops. Analysts can thus identify patterns and make predictions about the type of attacks that are most likely to occur, the sorts of entities that are most likely to be targeted, and the types of actors most likely to engage in these behaviors.

Cybersecurity data science is a field of study oriented toward the study of security incident patterns; it aims to make cybersecurity automated and intelligent (Sarker et al., 2020). Cybersecurity data scientists use existing cybersecurity incident datasets, including historical or time-series data, to train machine learning models to identify, forecast, and respond to security breaches.

This new and exciting field of study is concerned with three ethical values or goals. First, its practitioners are concerned with confidentiality, or the ability to prevent unauthorized individuals from accessing or disclosing information. Next, they are concerned with the integrity of information by preventing intruders from destroying, modifying, or releasing information in ways that can harm users or society. Finally, they are concerned with efficiently making information available to authorized users while preserving privacy.

Specialists examine how artificial intelligence can create better security policy rules – regulating how users may or may not access data, deciding how breaches or unauthorized requests will be managed, and generating security alerts.

Applying the Lenses

Thus far, we have looked at specific problems – such as the rights and responsibilities of data producers and those who work with data, the ethics of using biometric data, and problems related to connectivity. In our final section of this chapter, we will apply the three specific ethical lenses – virtue ethics, utilitarian ethics, and deontological ethics – to consider the more general issues of privacy, spying, and surveillance.

Virtue Ethics

In 1929, US Secretary of State Henry Stimson decided to close the State Department's codebreaking office, believing that the activities of spying and invading other's privacy were not in keeping with what he saw as the diplomatic mission of the State Department. Later, in the aftermath of the bombing of Pearl Harbor, he was asked to explain how he had decided for the State Department to stop engaging in espionage activities. In response, he famously stated that "gentlemen do not read each other's mail" (Burtness & Ober, 1968, p. 27).

We can view this statement as a summation of the virtue ethics position on privacy. Stimson referred to breeding and character in justifying his decision to end the State Department's involvement in espionage activities and what he saw as a "gentleman's" orientation towards the rights of others, including the right to privacy. Here Stimson could be seen as supporting the virtue of *restraint* – in acknowledging that just because one can do something, one can still decide to discipline oneself not to do it. As Steele points out in his discussion of when states should engage in restraint, Aristotle first asked questions regarding the conditions under which a person should be prudent or continent and the conditions under which one was likely to be incontinent or unrestrained. Aristotle argues that there are many situations in which one might find oneself carried forward by desire or want, but that restraint refers to the quality of stopping this momentum (Steele, 2014, p. 9). Steele notes that an individual might wish to do something due to social conditioning or peer pressure and that restraint may thus require strength to resist these forces. Many professions include restraint as a professional ethic. As noted at the beginning of this chapter, physicians, psychologists, and even priests may be in a position whereby they are provided with access to private information. However, their professional training emphasizes that they are morally required to restrain themselves from sharing or profiting from that information. Similarly, Black (1994) argues that journalists frequently engage in restraint in, for example, declining to publish the names of rape victims, even if they could profit by providing the public with this information.

Virtue ethics thus suggests that even if one has a solid desire to invade another's privacy, even if the environment is structured so that one could get away with doing so, and even if one is pressured to do so, the virtuous individual (or organization or state) should choose to improve his character by engaging in restraint. The shorthand statement "gentlemen do not read each other's mail" is also absolute. It is unambiguous, and there is no room in the statement for compromise – such as allowing a limited amount of spying, invasion of privacy, or allowing it in a limited range of circumstances.

For those who regarded Stimson and his decision as responsible for the mass casualties sustained in the Japanese attack on Pearl Harbor, the virtue ethics position seemed naïve and idealistic and not in keeping with the realities of a world that was on the brink of war. Considering their objections to Stimson's argument, we can consider why people might object to a virtue ethics argument against spying or espionage. First, those who disagree with Stimson point out that Stimson, in stating that "gentlemen do not read each other's mail," was not simply stating his position but instead was assuming that there was a norm or shared understanding amongst all who engaged in diplomacy. That is, a decision not to spy because doing so was not in keeping with one's character as diplomacy would work only if everyone engaged in diplomacy behaved in the same way. Otherwise, the gentleman merely stands to be taken advantage of (or exploited and defeated) by others who are less gentlemanly. We might point to the actions of other nations – like North Korea or Russia. They are willing to "read each other's mail" or engage in espionage to argue that the United States could not afford to take the virtue ethics position without conceding defeat to less virtuous opponents.

Indeed, in Plato's *Republic*, Plato recounts the story of a shepherd, Gyges, who receives a ring that can render him invisible (and thus capable of spying). Here, Plato asks us to consider whether anyone, given such a power, would be able to restrain himself, only using the power for good. Plato suggests that inevitably, one might decide to use this power to gain an unfair advantage over one's companions. In the story, Gyges uses his power of invisibility to seduce the king's wife and gain riches and power. Plato's story suggests that one could be a "virtuous spy" if one could use these powers with the right intent, but acknowledges that the temptation to use the same power for less-than-virtuous intents is powerful (Allen, 2008).

Similarly, Stimson's detractors suggested that he was naïve in assuming that all diplomats would have the same intent, deciding to eschew spying to respect others' rights. However, in later years, Stimson became the head of the US Department of War. In this capacity, he supported the use of espionage by the military community. Thus, one might argue that Stimson's original position was naïve since he could only be moral and virtuous as a diplomat because individuals in other groups – such as the military – were willing to violate their character and principles to defend US national interests. The practical ethics of others allowed Stimson to maintain his position on virtue ethics.

Utilitarian Ethics

As noted, virtue ethics focuses on character. In contrast, practical ethics asks: What type of good is produced by engaging in surveillance and spying in this situation? Does the utility produce through engaging in privacy violations outweigh any potential harm from violating someone's privacy?

The most famous practical statement on spying is the sentence uttered by the American soldier Nathan Hale, who collected intelligence about British activities during the Revolutionary War. (He was later captured and executed by the British.) Hale famously justified his espionage activities and the deception which they required by stating that

"Every kind of service necessary to the public good becomes honorable by being necessary" (quoted in Robarge, 2007).

More recently, journalist Olga Khazan (2013) defended US and British surveillance practices, arguing that the first goal of public servants should be to advance a country's interests and keep its citizens safe. She describes how the British intelligence agency GCHQ set up fake internet cafes for delegates during the 2009 G20 meetings in London. The computers provided software that logged users' keystrokes and recorded their communications. Khazan (2013) argues that a nation concerned with cyberwarfare and terrorism threats is justified in engaging in codebreaking, intercepting mail, and telegraph communications.

Corporations have made similar utilitarian arguments to explain why they monitor their employee's activities online. Covert suggests that such activities are a necessary evil, allowing companies to continue to operate as productive, efficient corporations, thereby enriching shareholders and providing employment for their workers.

Deontological Ethics

Finally, let us consider the deontological approach. Remember, this approach includes an injunction against treating people as a means to an end. Deontologists believe that it is unethical to manipulate or deceive people to get access to their data. It would also be inappropriate to treat people merely as sources of data.

Thus, deontologists worry about issues of *gatekeeping*, where people are required to provide personal information like an e-mail address or access to their Facebook account to read an article or visit a website. Here users have little choice and no autonomy in deciding whether or not to provide their data. Indeed, one could argue that they are being coerced into providing this information (Alterman, 2003). Bergstrom (2015) also expresses concern about situations in which users are promised a reward for furnishing personal data, such as a faster download speed for an article or additional points in an online game. He asks if individuals are being manipulated or misled into furnishing personal data.

Thus, to treat users with dignity and respect, states and corporations must have strict procedures regarding informed consent (Culnan & Williams, 2009). They also need to consider questions of equity by asking: Does everyone have an equal ability to refuse to provide personal information, or are those who are less literate or less wealthy at a disadvantage? Does everyone understand equally well what they agree to?

However, not everyone accepts this approach. Some critics say deontologists focus too much on individual rights while neglecting how communities often benefit through large-scale data collection. For example, we can consider practices like tracking the spread of infectious diseases by collecting information on internet searches for flu symptoms. Here a utilitarian would argue that preventing the deaths of vulnerable populations – like infants and the elderly – through engaging in health surveillance is a greater good than protecting the rights of individuals to be free from such surveillance.

CHAPTER SUMMARY

- Decision-makers increasingly face trade-offs between an individual's right to keep his private life private and the obligations of employers and the government to keep citizens safe.

- Privacy's understanding evolves with new technologies that blur boundaries between public and private space and public and private life.
- It is often unclear who your data belongs to – When does it cease being yours, who has the right to collect it, and what are their obligations to safeguard it?
- Some analysts argue that individuals have an ethical responsibility to share their data to achieve large-scale social goals, like responding to a pandemic.

DISCUSSION QUESTIONS

1 **Designing for Privacy**

You are a software designer, and you are asked to create a phone app that includes a health contact tracing app as part of the phone's programming. The contact tracing app cannot be removed from the phone's programming but will be activated and visible to users only if a new pandemic occurs. The national health authorities deem it necessary. The app will rely on DIRECT contract tracing, and data collection will be centralized.

- What are some of the ethical issues that you can identify in such a request?
- What ethical principles might you refer to in accepting this request?
- What ethical principles might you refer to in refusing this request?
- Can you think of any modifications you might wish to make to data collection or storage procedures that might make you more comfortable ethically with this request?

Alas, you did not get the contract to develop the app, but someone else did, and the project went forward. This app now comes preloaded on all the phones in your country. But you are an excessively skilled hacker, and if anyone can work with and modify this app, it would be you! Three customers come to your shop seeking your expertise.

Customer A comes to your shop and asks you to remove the contact tracing app from his phone. Do you do so? Why or why not? Would you regard doing so as illegal? Unethical?

Customer B asks if you can decrypt the data sent to the government data storage facility. He asks you to write him a program to do so but won't tell you what he needs it for. Do you help him? Why or why not? Would you regard doing so as illegal or unethical?

Customer C asks if you can write a program that will allow him to fabricate new user data and insert it into the data storage site. He also won't tell you who he is and why he needs to do this. Do you help him? Why or why not? Would you regard doing so as illegal or unethical?

Do you report any of these requests to law enforcement?

Are the three requests detailed here the same or different regarding the ethical issues raised? How do they differ, and how are they the same?

2 **Emerging Technology and Sentiment Analysis**

Many ethicists are concerned about how emerging technology in sentiment analysis might be deployed in the future. They suggest that next-generation biometrics can tell who you are and how you are feeling or your state of mind. Royakkers et al. (2018) state that mental privacy should be absolute: "We are talking about people's right and ability to keep private what they think and feel."

Imagine the following scenario: You are being interviewed for a job over Zoom, and you are asked, "Do you support the mission of this organization?" The gathered biometric data indicates another candidate for the job, who is more loyal and more supportive of the organization's mission – based on facial recognition and an audit of your feelings towards the organization.

Should an employer be able to command your loyalty in this way? What if you had been working there five years, and they decided to fire you based on an analysis of your facial expressions, which suggested that you didn't support the organizational mission? Should that be allowed?

What if it was a national security position rather than selling automobiles? Would that change your calculus?

You can use a values-based and a design-based viewpoint in your responses.

3 **Thinking About Workplace Privacy**
   You are offered a position in your field that pays twice your current salary, but it comes with the caveat that a microchip will be installed in your brain. You will be subject to remote surreptitious brain surveillance. Would you take the job? Why or why not?

RECOMMENDED RESOURCES

Abbiati, G., Ranise, S., Schizzerotto, A., & Siena, A. (2021). Merging datasets of cybersecurity incidents for fun and insight. *Frontiers in Big Data*. https://doi.org/10.3389/fdata.2020.521132. Retrieved from https://www.frontiersin.org/articles/10.3389/fdata.2020.521132/full.

Haselager, P. (n.d.). Brain reading and mental privacy. *TED Talks*. Approximately 13 minutes. Retrieved June 22, 2021, from www.ted.com/talks/pim_haselager_brain_reading_and_mental_privacy.

Solove, D. (2021). Teach privacy training blog. The website Teachprivacy.com contains information on privacy laws in most nations of the world and has a great collection of links to other privacy resources. It may be accessed at https://teachprivacy.com/privacy-security-training-blog/.

Stanley, J. (2019). *Should you buy a ring doorbell camera?* Las Vegas, NV: American Civil Liberties Union. Retrieved June 21, 2022, from www.aclunv.org/en/news/should-you-buy-ring-doorbell-camera.

YouTube tutorial on how to obfuscate your data so it cannot be scraped (approximately 15 minutes): John Watson Rooney. November 5, 2021. Hiding data with Javascript? Web scraping obfuscation. *YouTube*. Retrieved June 21, 2022, from www.youtube.com/watch?v=ks-iekIJy6M.

References

Abbiati, G., Ranise, S., Shizzerotto, A., & Siena, A. (2021, January). Merging datasets of cyber security incidents for fun and insight. *Frontiers in Big Data*, *3*(1). Retrieved June 1, 2022, from www.ncbi.nlm.nih.gov/pmc/articles/PMC7931890/#:~:text=TABLE%201%20%20%20%20Id%20%20,%204.%20Government%20%2023%20more%20rows%20.

Aicardi, C. D. (2016). Emerging ethical issues regarding digital health data: On the World Medical Association draft declaration on ethical considerations regarding health databases and biobanks. *Croatian Medical Journal*, *57*(2), 207–213.

Allen, A. L. (2008, January). The virtuous spy: Privacy as an ethical limit. *The Monist, 1*(1), 3–22.

Al-Saggaf, Y., & Islam, M. Z. (2015). Data mining and privacy of social network sites users: Implications of the data mining problem. *Science and Engineering Ethics, 21*(4), 941–966.

Alterman, A. (2003). A piece of yourself: Ethical issues in biometric identification. *Ethics and Information Technology, 5*(3), 139–150.

Amoore, L. (2021). Deep borders. *Political Geography*. https://doi.org/10.1016/j.polgeo.2021.102547.

Belloti, V. (1997). Design for privacy in multimedia computing and communications environments. In P. E. Agre (Ed.), *Technology and privacy: The new landscape*. Boston, MA: MIT Press.

Bergstrom, A. (2015). Online privacy concerns: A broad approach to understanding the concerns of different groups for different uses. *Computers in Human Behavior, 53*, 419–426.

Black, J. (1994). Privacy in America: The Frontier of duty and restraint. *Journal of Mass Media Ethics, 9*(4), 213–234.

Bok, S. (1989). *Secrets: On the ethics of concealment and revelation*. New York: Vintage Books.

Boyd, D., & Marwick, A. (2011). Social privacy in networked publics: Teens' attitudes, practices and strategies. *Apophenia Blog*. Retrieved June 2, 2022, from www.zephoria.org/thoughts/archives/05/09/2011/how-teens-understand-privacy.html.

Brey, P. (2007). Ethical aspects of information security and privacy. In N. A. Petkovic (Ed.), *Security, privacy and trust in modern data management* (pp. 21–36). Heidelberg; Berlin: Springer.

Burtness, P., & Ober, W. (1968). Secretary Stimson and the First Pearl Harbor Investigation. *Australian Journal of Politics and History, 14*(1), 24–36.

Cavoukian, A. C., Chibba, M., & Stoianov, A. (2012). Advances in biometric encryption: Taking privacy by design from academic research to deployment. *Review of Policy Research, 29*(1), 37–61.

Chinchilla, R. (2012). Ethical and social consequences of biometric technologies. In *American society for engineering education annual conference*. Retrieved June 2, 2022, from www.asee.org/public/conferences/8/papers/3789/view.

Culnan, M., & Williams, C. C. (2009). How ethics can enhance organizational privacy: Lessons from the choice point and TJX data breaches. *MIS Quarterly, 33*(4), 673–687.

Davis, K. (2012). *Ethics of big data*. Sebastopol, CA: O'Reilly.

Duvall, N. M., Copoulos, A., & Serin, R. C. (2016). Comprehension of online informed consent: can it be improved? *Ethics and Behavior, 26*(3), 177–193.

Ellerbrok, A. (2011). Playful biometrics: Controversial technology through the lens of play. *The Sociological Quarterly, 52*(4), 528–547.

Etzioni, A. (2015). A cyber age privacy doctrine: More coherent, less subjective and operational. *Brooklyn Law Review, 80*(4), 1267–1296.

Floridi, L. (2013). *The ethics of information*. Oxford: Oxford University Press.

Giang, V. (2013, July 3). Eleven common interview questions that are actually illegal. *Business Insider*. Retrieved June 2, 2022, from www.businessinsider.com/11-illegal-interview-questions-2013/.

Hoepman, J.-H. (2013). Privacy design strategies. *International Federation for Information Processing*. Retrieved June 1, 2022, from https://sites.law.berkeley.edu/privacylaw/2013/10/03/jaap-henk-hoepman-privacy-design-strategies/.

Khazan, O. (2013, June 17). Gentlemen reading each other's mail: A brief history of diplomatic spying. *The Atlantic*. Retrieved June 2, 2022, from www.theatlantic.com/international/archive/2013/06/gentlemen-reading-each-others-mail-a-brief-history-of-diplomatic-spying/276940/.

Kligiene, S. (2012). Digital footprints in the context of professional ethics. *Informatics and Education, 11*(1), 65–79.

Marmor, A. (2015). What is the right to privacy? *Philosophy and Public Affairs, 14*(13), 3–26.

Metcalf, J. (2014). *Ethics codes: History, context and challenges*. Retrieved June 2, 2022, from http://bdes.datasociety.net/council-output/ethics-codes-history-context-and-challenges/.

Michmerhuizen, S. (2007). *Confidentiality, privilege: A basic value in two different applications*. Washington, DC: American Bar Association Center for Professional Responsibility.

Moor, J. (1997). Toward a theory of privacy in the information age. *SIGCAS Computers and Society, 27*(3), 27–32.

Moore, A. D. (2005). Introduction to information ethics: Privacy, property and power. In A. D. Moore (Ed.), *Information ethics: Privacy, property and power.* Seattle, WA: University of Washington Press.
National Information Standards Organization. (2004). *Understanding metadata*. Bethesda, MD: NISO Press.
Plato. (1967). *The laws*. Cambridge, MA: LCL.
Rasner, G. (2020). *Cybersecurity and third party risk*. Hoboken, NJ: Wiley.
Robarge, D. (2007). *A review of fair play: The moral dilemma of spying*. McLean, VA: Central Intelligence Agency Library Center for the Study of Intelligence.
Royakkers, L., Timmer, J., Koo, L., & van Est, R. (2018). Societal and ethical issues of digitization. *Ethics and Information Technology*, *20*(2), 127–142.
Sarker, I., Kayes, A., Badsha, S., Alqahtani, H., Watters, P., & Ng, A. (2020). Cybersecurity data science: An overview from a machine learning perspective. *Journal of Big Data*, 7(41). https://doi.org/10.1186/s40537-020-00318-5.
Schneier, B. (2017, July 18). Surveillance is the busness model of the internet. *Schneier on Security*. Retrieved June 1, 2022, from https://doi.org/10.1186/s40537-020-00318-5.
Sims, J. (2010). A brief review of the Belmont report. *Dimensions of Critical Care Nursing*, *29*(4), 173–174.
Sloan, J., Cherel, C., & Yang, A. (2012). Social media and the public workplace: How Facebook and other social tools are affecting public employment. *League of California cities annual conference*. Retrieved January 2, 2017, from www.publiclawgroup.com/wp-content/uploads/2012/12/2012-09-01/League-of-CA-CitiesSloanYang-Social-Media-and-the-Public-Workplace-FINAL.pdf.
Sloan, R. H., & Warner, R. (2014). *Unauthorized access: The crisis in online privacy and security*. New York: CRC Press.
Steele, B. (2014). *Of body and mind, of agents and structures: Akrasia and the politics of restraint*. Paper presented at the 2014 ISA-West Meeting, Pasadena, CA. Retrieved January 3, 2017, from www.academia.edu/15574573/Akrasia_and-Restraint_in-International_Relations.
Tavani, H. (2008). Floridi's ontological theory of information privacy: Some implications and challenges. *Ethics and Information Technology*, *10*(1), 155–166.
United States Department of Health and Human Services. (2013). *Considerations and recommendations concerning internet research and human subjects research regulations, with revisions*. Retrieved January 10, 2017, from www.hhs.gov/ohrp/sites/default/files/ohrp/sachrp/mtgings/2013%20March%20Mtg/internet_research.pdf.
Van den Hoven, N. J. (1997). Privacy and varieties of moral wrong-doing in an information age. *SIGCAS Computers and Society*, *27*(3), 29–37.
van Dijk, N., Tanas, A., Rommetveit, & and Raab, C. (2018). Right engineering? The redesign of privacy and personal data protections. *International Review of Law, Computers and Technology*, *32*(2–3), 230–256.
Virginia poised to enact comprehensive consumer privacy law. *JDSUPRA.com*. February 17, 2021. Retrieved from https://www.jdsupra.com/legalnews/virginia-poised-to-enact-comprehensive-6293368/.
Vitak, J., & Zimmer, M. (2020). More than just privacy: Using contextual integrity to evaluate the long-term risks from COVID-19 surveillance technologies. *Social Media and Society*, 1–4.
Warren, D., & Brandeis, L. (1890). The right to privacy. *Harvard Law Review*, *4*(5), 193–220.
Wilkins, L. (2008). *Handbook of mass media ethics*. New York: Routledge.
Wu, X., Wu, T., Khan, M., Ni, Q., & Dou, W. (2021). Game theory-based correlated privacy preserving analysis in big data. *IEEE Transactions on Big Data*, 7(4), 643–656.

5 The Problem of Surveillance

LEARNING OBJECTIVES

At the end of this chapter, students will be able to:

1. Define different types of surveillance (covert, asymmetrical, differential)
2. Describe laws that affect surveillance practices
3. Articulate a virtue ethics, consequentialist/utilitarian, and deontological argument in favor of and against surveillance
4. Describe the trade-offs between privacy, surveillance, and security

As we consider the ethical problem of surveillance in this chapter, we can consider four real-life situations involving the use of surveillance by a government, corporation, or individual:

- In the fall of 2016, US intelligence agencies expressed concerns regarding foreign interference in the US presidential election process. In the summer of 2016, they noted that Russian intelligence agents had perhaps compromised the integrity of the US election process by hacking into voter databases in Illinois and Arizona and making public the contents of e-mails sent by presidential candidate Hillary Clinton and members of the Democratic National Committee (Fritcke, 2016).
- In the United States, many employees who work off-site may have their activities monitored by their employers through GPS tracking in a company vehicle or company-provided phone. Employees are often monitored during working hours and when driving or using the phone. Recently, reporter Kaveh Waddell (2017, p. 1) asked readers to consider the following: "Someone's tracking your phone or has placed a GPS device on your car. They know if you're at home, a pregnancy clinic, church, or a gay bar." Currently, there is no federal-level legislation regarding the rights of employees, and decisions as to who constitutes a violation are often made on a case-by-case or state-by-state basis.
- In recent years, the Dutch government has encouraged citizens to engage in *participatory surveillance*. Citizens who find themselves in an emergency – such as a natural disaster or terrorist event – are encouraged to use their phones to record videos of the scene to aid law enforcement later in finding the perpetrators. The Dutch government defines participatory surveillance as a logical part of responsible citizenship (Timan & Albrechtslund, 2015).
- In January 2022, multiple women posted videos on the video-sharing site TikTok. They described their dating experiences with a man named Caleb, who claimed

DOI:10.4324/9781003248828-7

to work for furniture manufacturer West Elm. Several women described how they matched with him on a dating platform, spent time with him on dates where he wooed them, and proceeded not to return their texts and messages. The women warned others not to date Caleb, describing him as a "serial dater" (todayheadline.co).

What do these stories show? First, they illustrate that surveillance encompasses various activities carried out by various actors – corporations, nation-states, nonstate actors, and private citizens. Surveillance activities may be illegal or legal or fall into a grey area where they are poorly regulated and understood. In the United States, legal scholars often use the term **surveillance** to refer only to the *unauthorized* collection of personal or professional information and associated activities such as the unauthorized publication or sharing of that information.

Surveillance, however, is often taken to mean collecting information without informed consent. Under US law, an entity (such as the government or a corporation) is engaged in electronic surveillance if they acquire wire or radio communications – through Medtronic, mechanical, or another surveillance device. The entity must collect information sent by or received by an American in the United States – when the individual could have a reasonable expectation of privacy and where one would typically require a warrant to collect this information for law enforcement purposes. It is also considered electronic surveillance if an entity installs a device on your computer or another device to collect information – in a circumstance where you could expect privacy or where a warrant would be required to collect this information for law enforcement purposes (Chapter 50 of US Criminal Code, Section 1801). US law explains the circumstances under which it is illegal to engage in electronic surveillance and the conditions under which it is legal (such as when one is suspected of plotting to harm the United States and the proper authorities have given the entity permission to collect this information). This definition assumes that the individual or group that is the subject of surveillance is unaware that they are being surveilled and have not been informed or warned. They have thus not consented to the surveillance.

We can also distinguish between *detection from a distance* and *obtrusive detection*. Detection from a distance might include a drone collecting information from the sky, a spy satellite, or closed-circuit television systems (CCTV), which record information and engage in specified activities such as identifying license plate numbers. Let's think back to the example of the employee whose employer uses a GPS to track their activities. This is an example of detection from a distance – since the employee is unaware he's being tracked. In contrast, intrusive detection may include a scanner (such as at the airport), biometrics, or access control cards. Here, the individual knows that his information is being collected and consents to surveillance.

Second, these stories show that the people and groups being watched might not yet have committed a crime or experienced a negative outcome. Instead, surveillance activities may include using analytics to *preemptively* detect patterns of inquiry or activity, leading to future harm. Then, an agency can take steps to preempt that negative outcome. Here, *analytics* is defined as "the use of information technology to harness statistics, algorithms and other tools of mathematics to improve decision-making" (Schwartz, n.d., p. 2) We can see how health care personnel use analytics in conducting health surveillance. Public health officials can preemptively monitor disease patterns – such as beginning a vaccination campaign – to prevent an epidemic (Longjohn, 2009).

For this reason, Agre warned in 1994 that surveillance might have two components in the future. Those engaged in surveillance might be faulted ethically for watching you without your knowledge or permission (spying). Still, they might also be faulted ethically for engaging in the "capture" of your data. He warns that once a data architect has captured your data, they acquire the power to interact with it in various ways. Ultimately, Agre argues, those who design interfaces and programs can now "steer" you towards engaging in some activities or away from other activities – based on what they know about you because they have succeeded in capturing your data. While, at first glance, such steering activities may seem rather mundane (i.e., sending someone who has "liked" a picture of a cruise ship on Facebook an offer for a reduced price on a cruise), Agre worries that surveillance can lead to data capture, which can ultimately lead to a waning of individual autonomy as we are "steered" towards performing specific actions, often without our full knowledge and consent (Agre, 1994).

Third, these stories show that not all surveillance is equally intrusive. They show that one can conduct ethical surveillance – by setting limits on the type of information collected, establishing norms for the storage and sharing of this information, and establishing rules for respecting the rights of those subject to surveillance.

Fourth, these stories show that those who engage in surveillance may have various motives – some of which might be ethically justified, while others cannot. For example, public health officials are protecting citizens' health and security; therefore, monitoring – even in situations that might be intrusive by the subjects – is necessary to provide health security to society. Similarly, computer scientists use algorithms to monitor employee behavior within an organization to predict who is likely to be an *insider threat*. Here, certain types of activities suggest that someone might be a malicious insider – such as patterns where they are accessing data that they have no job-related need for, erasing logs of their attempts to access information, or their use of programs to map internal networks or identify weaknesses in a system (Azaria, 2007).

Finally, these stories raise questions regarding the responsibilities of those who produce, provide, and use surveillance technologies. To what degree should they be held ethically responsible if others misuse their created technologies? In the wake of the 2013 Edward Snowden revelations, analysts asked whether internet sites like Google and Facebook knew that US government agencies were conducting surveillance of individuals who visited those sites. They suggested that platforms like Google and Facebook bore some ethical responsibility for the privacy abuses. These platforms should have considered how the available information might be used against citizens.

Box 5.1 Tech Talk: What Is Geofencing?

The term geofencing refers to the process of creating a virtual geographic boundary around an area using GPS or RFID technology. A program may monitor an area, triggering an alert when the area is entered, or collecting data on those who enter an area.

The term *geofencing* is most often associated with targeted marketing. For example, those individuals in the vicinity of a courthouse might receive a targeted announcement about the availability of legal services from a law firm. However, geofencing might also be used to send information to a jury about a case. For example, in a product liability case, the company serving as a defendant might send information about their product's safety to everyone in the courthouse, hoping jurors would read it and be influenced (Downey, 2019).

Political advertisements are often highly targeted based on geofencing as well. Specific demographics and zip codes might receive political information considered to be most relevant and salient to them.

Is Geofencing Legal?

Currently, at least in the United States, information sent via geofencing is considered a protected form of freedom of speech. However, at this time, US legal scholars have raised issues regarding reverse geofencing. Reverse geofencing refers to a procedure where a group like a law enforcement agency might request information regarding who was in a specific geographic area to gather evidence or identify a suspect in a break-in. It is unclear whether companies have a legal or an ethical obligation to comply with reverse geofencing requests (for example, a request to Google for user phone GPS data tracking) (Cheung, 2021).

However, the European Union's General Data Protection Regulation and the California Consumer Privacy Act appear to conflict with current geofencing practices for social marketing purposes. Here, the EU GDPR requires that people consciously opt to collect their personally identifiable data, which does not occur in situations of passive geofencing.

Bibliography

Cheung, K. (2021, November 21). Abortion in the surveillance state. *Jezebel.com*. Retrieved from https://www.msn.com/en-us/news/us/abortion-in-the-surveillance-state/ar-AAQZSnZ/.

Downey, M. (2019). Legal ethics and geofencing. *Litigation*, 64.

Local, D. (2020, January 13). Ethics in geofencing: Post GDPR and CCPA. *Demand Local*. Retrieved from https://www.demand.local.com/blog/ethics-in-geofencing/.

Surveillance in History

Although the examples which began our discussion are all from the twenty-first century, surveillance is hardly a new idea. Indeed, as Stoddart points out, most major religions have a concept of a God who watches over his people. Such a God is omniscient, all-knowing, omnipotent, or all-powerful. Stoddart argues that historically, people have been comforted by this notion, counting on their God to provide them with security and justice – believing that their God sees who is behaving unethically and perhaps even intervenes to punish the unjust and reward the just. Thus, we have always had an idea of benevolent surveillance rather than regarding surveillance as inherently "creepy" (Stoddart, 2011).

In addition, we can trace the field of public health back to the Middle Ages. Throughout the late 1400s into the eighteenth century, communities collected information or intelligence on the spread of plague to prepare and hopefully save their villages from destruction. Officials established quarantine zones based on observing patterns of disease movement or health surveillance. The goal here was to stop the spread of disease by establishing procedures for monitoring its transmission and lethality.

Finally, we can point to the experiments by Jeremy Bentham, a British social reformer. In the late 1780s, Bentham developed a plan for a prison called the "panopticon." As McMullan describes the arrangement:

> The basic setup of Bentham's panopticon is this: there is a central tower surrounded by cells. In the central tower is the watchman. In the cells are prisoners . . . The tower shines bright light so that the watchman can see everyone in the cells. However, the people in the cells can't see the watchman and, therefore, assume they are always under observation.
>
> (McMullan, 2015)

Here we should remember that Bentham was one of the first utilitarian philosophers. He supported this form of surveillance because it was a highly efficient use of resources. One watchman could watch a larger population and preempt problems before they occurred, thus conserving resources. Surveillance allowed for reducing risk and uncertainty and thus reduced the likelihood of loss or damage. In addition, Bentham argued that not knowing whether or not they were under surveillance at any given time, the prisoners would begin to self-police their behavior, thinking twice before engaging in forbidden activities since they never knew if they were being watched. Self-policing would allow for the expenditure of fewer resources to watch the prisoners.

The Legal Climate

We can formulate arguments both in favor of and against surveillance. Some of the strongest arguments against surveillance come from the American and international legal communities, which have suggested that surveillance is often illegal and unconstitutional.

Here we should note that government surveillance of citizens – and legal opposition to this surveillance – did not begin with 9/11. After World War II, the US National Security Agency began reading international corporate communications sent by telegram. Both the Central Intelligence Agency and the NSA opened and read the mail of individuals and groups during this time frame. And as the United States learned during the McCarthy hearings, US intelligence had also surveilled US citizens (including celebrities) suspected of being Communist agents throughout the 1950s (Donahue, 2012).

In 1975, the US Congress conducted the Church Committee Hearings, looking into what they saw as overreach by the US intelligence community, in the aftermath of the Watergate scandal. (The Watergate scandal concerned activities by the Republican National Committee, which was interested in re-electing Richard Nixon, by collecting information on his opponents through illegal wiretapping of their election headquarters' phones.) The mandate of the intelligence community was to engage in the monitoring of foreign threats – but the Church Committee found that the intelligence community had exceeded its bounds, also monitoring the communications of Americans within the United States, and that it had not been sufficiently attentive to the Constitution and the system of checks and balances described in the Constitution.

These hearings led to the passage of the *Foreign Intelligence Surveillance Act of 1978*. This act spelled out the specific circumstances under which the US government could monitor communications between US citizens and people residing within the United States.

It placed the intelligence community's activities under both legislative and judicial oversight. In cases where an American citizen's communications would be collected, those collecting the information were required to get a warrant issued by a court authorizing the surveillance.

The Foreign Intelligence Surveillance Act was updated and amended several times as technologies and threats facing the United States changed. In 2007, an amendment known as the Protect America Act was added in response to increasing al Qaeda threats. This amendment allowed the US government to increase surveillance of US persons outside the United States under certain circumstances. The amendment established that a citizen abroad might not have the same expectation of privacy as a citizen in the United States (Bazan, 2008, p. 10).

Today, legal analysts are attempting to resolve questions of *jurisdiction*, or whose laws should apply in a globalized world where data storage and access patterns may not always neatly align with geographical boundaries. As Cordero has pointed out, some nations have stringent laws regarding the situations in which an ISP must be required to provide data to local or national authorities, some have lax laws, and some have none. She argues that while some people find US laws problematic, the bigger problem is that nations with no legal framework governing surveillance may still be engaged in surveillance, creating a situation where the citizen has few rights to object or even to know surveillance is taking place.

Legal analysts also worry that a French citizen might perform a search using Google, which seems suspicious to American authorities. Some analysts would say that the data here is French since the searcher was a French person performing a search on her home computer located in France. However, others might say that the data is American since it is housed on a Google server in the United States. The question is, thus, whose laws would apply? Is the French citizen protected from having to share her data with the NSA, for example, under French law – or does American law supersede French law if the data is American?

Within international law, states sign Mutual Legal Assistance Treaties (MLATs), which allow them to ask each other for assistance in criminal prosecutions. Countries can use MLATs to ask other countries to share data in the cyber arena. However, this is a lengthy, bureaucratic process. The United States has argued that in prosecuting cases like terrorism, it takes too long for the other country to get a warrant and carry out the necessary steps to provide the data. Thus, in April 2014, a New York district court ruled that Microsoft was obligated to turn over data – even though it was stored in Ireland – because Microsoft was incorporated in the United States under US laws. However, in other Supreme Court cases, the ruling has been that the United States does not have the right to seize property outside of the country.

In November 2015, Microsoft unveiled a new plan. Under this plan, foreign users can store their data at a center located in Germany, operated by a German provider. This is referred to as **data localization**. The German provider is referred to as a **data trustee**. Under local laws, this entity then decides who may access the data and under what circumstances. As Basu argues, "for this trustee model to work, the country in which the data is being hosted must have a robust data protection framework within its domestic law" (Basu, 2015, p. 1). While this solution seems to provide increased privacy rights to users, opponents claim it destroys the flexibility of the internet, where data can and should be duplicated worldwide for backup and efficient access. People are afraid that localization will lead to balkanization, or the creation of separate national internets with different rules governing them, and a slowing of the efficiency in which data currently zooms worldwide.

Currently, legal analysts disagree about whether different countries should have different standards and laws regarding how much encryption to allow, how much surveillance to allow, and how a nation's government can behave in an extraordinary situation (i.e., whether a national government should be allowed to intervene to shut down the internet if it perceives some form of threat – like terrorism or outside manipulation of an event like a national election). As noted in Chapter 4, different societies may have different understandings of what constitutes an invasion of privacy or improper surveillance, with different standards prevailing in different nations and regions. What one nation regards as a good set of precautions to preserve civil rights may seem unduly restrictive to another nation's government, based on national understandings and the threat they face (Saran, 2016).

Ethical Critiques of Surveillance

In addition to legal critiques and challenges in surveillance, we can identify ethical critiques. As Buff has argued, data itself is neutral. However, the use of data is not ethically neutral. Managers and technicians who work in surveillance make ethical decisions regarding what data to collect, store, distribute, and use. They may do so based on a solid ethical code that emphasizes respect for the targets of surveillance, but the potential exists for abuse by those who engage in surveillance or watching.

The most common ethical criticisms of surveillance focus on how it establishes *power asymmetries* between the watcher and the watched; the ability of surveillance regimes to create an atmosphere of suspicion and mistrust in which people may have their rights to freedom of speech and assembly curtailed and actively engage in measures in which they curtail their rights to these goods; and the fact that surveillance is often applied differentially, with those who are most disenfranchised in society often becoming the subject of greater surveillance. As Michael and Michael (2011) write:

> The incongruity behind traditional surveillance technologies is that individuals of power and influence are not subjected to the extreme and exaggerated types of surveillance techniques designed and planned for everyone else.
>
> (p. 15)

As the example of Bentham's panopticon showed, surveillance technologies automatically set up a power asymmetry in which the watchers have a great deal of power, and the watched have very little. In such a situation, the watcher could easily abuse his power. Indeed, watching often conveys a sense of ownership over the subject and the subject's body. An abusive husband might hire a detective to watch his wife; pregnant women may find that strangers pay attention to smoking or drinking alcohol; celebrities may find that the general population pays attention to whether they are pregnant or merely getting fat. In countries with strict population policies, government agents have historically watched women to ensure they were not planning to abort a pregnancy (in Romania) or have a second child (in China).

While there are laws governing surveillance, there is no mechanism that assures that the watchers behave ethically 100 percent of the time and do not misuse this power, nor is there any requirement that the watchers treat those they watch with respect and empathy. Thus, critics often ask, "Who watches the watchers?" (Schneier, 2006).

At the same time, analysts worry that surveillance technologies will automatically lead to an expansion of state power. The American reformer Benjamin Franklin famously stated, "they that can give up essential liberty to obtain a little temporary safety deserve neither liberty nor safety." Therefore, individuals who support a limited role for government in society may be more suspicious of the benefits of surveillance and more attuned to its risks (Neocleos, 2003, p. 15).

In considering power asymmetries, we can also consider the question of equity and how those with better access to information could manipulate that situation to their advantage. Perlforth (2014) describes a situation that occurred on Wall Street when a group of former Morgan Stanley employees worked together to engage in unauthorized surveillance of company files. They used their access to gain insider information about several companies, which they then used to commit securities fraud. Although these employees didn't steal any substantial resources from their former employers, they might still be faulted for having committed an ethical breach – violating the principle of fairness. Scheppele (1993) writes that, "the hack was against the company, but the crime was actually against other investors" who did not have access to the same information in making their own investment decisions.

What Is a Surveillance Society?

Since the first edition of this book was written in 2017, surveillance has become even more widespread. Researchers estimate that approximately 200 million surveillance cameras are operating in China – or nearly one camera for every five people! But America is not that far behind. Fifty million surveillance cameras are currently operating in the United States, or nearly one for every six people!

And while previously, engaging in surveillance of one's employees, neighbors, or associates might have required a high level of technical skill, almost anyone can engage in surveillance today. Au (2021) describes the growing industry of "surveillance as a service" – noting that today one can subscribe to a service like the Ring doorbell and find oneself the owner of a complex system with a user-friendly interface that also comes with continual maintenance and updates and troubleshooting support. Today, **surveillance as a service (SaaS)** includes facial recognition and analysis, speech recognition and analysis, behavioral analysis, and nudging systems. In 2020, it was estimated that the facial recognition service industry was worth about $386 billion in the United States, while speech recognition services were worth $10 billion. Behavior analytics services are worth an estimated $890 million annually. And currently, there are few legal limits on using many of these services. In 2020, 109 countries approved facial recognition and analysis for government surveillance.

Today, surveillance programs can infer the identity of those recorded (through biometric surveillance and facial recognition), infer emotional states, and carry out lie detection through voice and facial analytics. The use of surveillance systems is particularly complicated from a legal and ethical perspective, since one nation may create programs that are then subscribed to and used by clients across the globe. For example, the French company Idemia currently offers surveillance solutions to clients in Bangladesh, Burkina Faso, Costa Rica, China, France, Germany, Kenya, Iceland, Italy, Mali, Norway, Singapore, and the United States. European Union draft regulations taking effect in 2023 regarding the use of artificial intelligence were expected to address some of the issues related to surveillance. However, as of this writing, it was unclear what effect this legislation would have on users in other nations.

Hartzog (2021) argues that most citizens currently benefit from privacy "obscurity." We often experience privacy simply because no one is very interested in looking at us. Here, he gives the example of someone walking across a room after taking a shower, wearing only a towel. They will simply be unobserved most of the time – by neighbors or someone walking past one's house. In the same way, you might assume that most of the time, unless you are a public figure or a celebrity, no one is particularly interested in observing you. However, Hartzog argues that our privacy as obscurity is vanishing with a growing number of cameras and facial recognition software. Technological advances increase the odds that our behaviors might be observed and linked to us through facial recognition. You might not remember who sat next to you on a plane six months ago, but facial recognition technology could remember and even store this information on your behalf.

As the costs of surveillance drop and surveillance procedures increasingly become automated, societal surveillance is likely to increase, leading some to conclude that someday privacy itself may be an old-fashioned concept.

Surveillance vs. "Spying": The Problem of Transparency

In thinking about surveillance, it is also essential to consider the value of **transparency**. As understood in a democratic society, transparency means that citizens have a right to know what their government is doing – the activities it is engaged in and the money it spends. Transparency is critical in providing *accountability*, or making governments (and corporations) accountable to people – their constituents or clients. Transparency is thus seen as a compelling way of combatting corruption or secrecy within government. Turilli and Floridi (2009, p. 107) describe information transparency as "enabling" other values, including accountability, safety, welfare, and informed consent. Companies must disclose problems with their products to keep their users safe, and governments must disclose information about their activities because they are accountable to citizens. They argue that governments do not have a right to keep secrets from citizens – except in dire situations such as wartime, where it might be dangerous to share information too widely. Furthermore, governments should act with the people's consent, and people cannot consent – or object – to something (like surveillance) when they are unaware that it is taking place.

The ethical value Edward Snowden defended in objecting to surveillance was transparency. In speaking to the British media in 2013 about the National Security Agency's surveillance program, he expressed concern that the government was undertaking surveillance of citizens without their awareness or consent. Furthermore, he expressed concern that the government was not disclosing information to citizens about these activities because they knew such actions were illegal and constitutional.

Today, similar ethical concerns have been raised regarding Big Data analytics. Turilli and Floridi describe the obligation of a company to inform users of a software package, for example, of what information about them and their use might be compiled as a result of their using the product, as well as the circumstances under which this information might be disclosed or shared and to whom.

Suspicion and Distrust

Other analysts ask what it does to people's sense of identity when they know they are being watched and even begin to alter their behavior in response to this knowledge (Vaz, 2003). Individuals who know they are the subject of surveillance can be said to have less freedom than others in society. They may also begin to think differently about themselves, no longer engaging in processes of self-expression and self-discovery, but rather being careful of how they and their behaviors are perceived – since their government may regard them not as innocent but rather as potentially disloyal. In a surveillance society, people may feel responsible for monitoring their behavior to avoid becoming a subject of suspicion.

The American Library Association has, for this reason, spoken out against unwarranted government surveillance, arguing that patrons should have the right to search for information without worrying about whether their searches will arouse suspicion. The American Library Association Code of Ethics (American Library Association, 2008) states, "We protect each library user's right to privacy and confidentiality concerning information sought or received and resources consulted, borrowed, acquired or transmitted."

Differential Surveillance

Critics also suggest that not everyone in society is equally likely to be the subject of surveillance. In the United States, many African American citizens refer to a phenomenon known as "driving while black," pointing out that an African American driver is stopped more often by the police, even when he has not broken any traffic rules. He is simply watched more closely because of his minority status in society. Those who receive government assistance sometimes note that when they shop at a grocery store, other customers may examine the contents of their shopping carts, wanting to make sure that tax dollars are not being wasted on frivolous food purchases.

Differential surveillance means that while surveillance claims to produce a collective good (such as security), not everyone pays the same for this collective good. For example, the Japanese Americans interned during World War II paid more than other Americans to preserve security. They were defined as potentially disloyal Americans; they sustained economic losses, lost employment opportunities, and separated from friends and family (Gillion, 2011). Further, one may argue that there is a mismatch between the individuals who probably should be most closely watched and those who are. Those with solid reasons to evade surveillance (such as those engaging in terrorist or criminal activities) may also have resources to enable them to do so (such as access to anonymous browsing software). Conversely, those with the fewest means to evade surveillance may become the target of increasing surveillance.

Use of Resources: Is Surveillance Wasteful?

Finally, today some critics are beginning to question the amounts of money, workforce, and resources the United States devotes to engaging in surveillance. They wonder if, given the relative infrequency of terrorist events, such resources couldn't be better spent on other activities of American civic life. In addition, one might ask whether a company should devote so many resources to identifying disgruntled company insiders, and whether those resources might be better spent learning what employees want and need and implementing programs that might keep them from becoming disgruntled.

Do We Have a Right Not to Be Surveilled?

In the aftermath of the Snowden revelations, several governmental bodies in the United States, Europe, and internationally spoke out against the phenomenon of unauthorized mass surveillance. In December 2013, the United Nations General Assembly suggested that the ability of governments to carry out large-scale surveillance could impact human rights. The US Privacy and Civil Liberties Oversight Board described National Security Agency (NSA) surveillance as violating the Electronic Communications Privacy Act and possibly the First and Fourth Amendments. The European Parliament's Committee on Civil Liberties, Justice and Home Affairs (LIBE Committee) cautioned that even in a situation where the monitoring seemed justified – such as in a fight against terrorism – there was no justification for "untargeted, secret or even illegal mass surveillance programs" (Amnesty International, n.d., p. 8).

An Amnesty International report entitled "Two Years after Snowden" also noted that 71 percent of respondents to a 15-nation poll about surveillance were strongly opposed to the government's spying on their citizens. Here, the report's authors noted that citizens might differ in the degree to which they oppose surveillance, reflecting different national norms. Nonetheless, no country voiced overwhelming support for such practices.

In 2018, the European Community adopted the *European General Data Privacy Regulation* (*GDPR*). This regulation states that European citizens have a right to data privacy. Their data can be collected only under specific conditions and for a legitimate purpose (i.e., city planning vs. prurient curiosity). The directive spells out how EU businesses and government entities must protect and store user data and the conditions under which it can be shared (Hawthorne, 2015). The regulation includes conditions for transferring EU citizens' data outside the EU, such as in the United States (European Commission, 2016). When citizens feel that their data privacy has been breached, they may claim financial compensation. They can also demand that their data be erased.

Box 5.2 Critical Issues: Ethics of Worker Surveillance

What do employees do while they are at work? A 2017 study suggested that in the United States, corporations lose approximately $15 billion worth of business each year due to employees wasting time on the job. The average worker spends five hours a week on non-work-related tasks while at work, including surfing the web and answering personal e-mails (Bresiger, 2017).

In response to this issue, workplaces in the United States and abroad have begun using Electronic Performance Monitoring (EPM) devices and programs to watch their employees. EPM technologies include keystroke trackers, programs that analyze the content of employee computer screens, and sophisticated algorithms to measure employee productivity by tracking everything from an employee's tone of voice during a meeting to the number of programs they initiate during a given year. During the COVID-19 pandemic, employers turned to EPM devices to watch employees working from home.

In addition, employers may use location trackers to monitor movements of employees who leave the office to make deliveries, meet with clients, or repair devices in a client's home. Finally, employees – from teachers to law enforcement officers to doctors in a hospital – might be asked to either wear a camera that records their activities, words, and movements or consent to work in a place where such cameras are regularly in use.

Ethical and Legal Issues

While these devices can improve employee productivity in the short run, what sorts of ethical and legal issues do they create? Do such employee monitoring practices violate employees' rights to privacy? The answer is that it depends.

One lens that can be applied in thinking about employee surveillance is the lens of consent and knowledge. Willis argues that employee surveillance practices are legal if employees consent to be tracked in this way and if they are aware that they are being tracked. However, other scholars argue that employees do not have the right to consent or negotiate with employers regarding whether or not they will be monitored since they are in a weaker position as job seekers who may need to earn a wage. And while theoretically, an employee who does not consent to monitor in one workplace may seek employment elsewhere, the growing use of such monitoring programs may mean that it is the norm in certain types of employment (like law enforcement or childcare). In some sectors, then, alternate employment without monitoring does not exist.

Do Employers Not Trust Their Workers?

A broader set of ethical arguments may be made by asking precisely what an employer is purchasing when they agree to employ you. Some scholars argue that an employer purchases your presence at the workplace for a set number of hours per week, particularly if you are paid an hourly wage. Thus, the employer is within his rights to stipulate that all of the time spent in the workplace during these hours should be spent working for the employer. When an employee works for a specific client, carrying out tasks like preparing a legal brief or filing their taxes, the client also has the right to expect not to be overcharged. EPM software can thus provide transparency, showing exactly how long an employee worked on a project and ensuring that the client is not billed for time not spent on the project (Willis, 2020).

But some employees feel that using EPM in a workplace means employers do not trust workers to make good decisions about how to complete their work and spend their time. Particularly in the post-COVID workplace environment, some employers have begun offering workers greater flexibility regarding how they spend their time or complete their tasks.

Different Regulatory Climates

As noted throughout this book, ethics and practices sometimes differ depending on the culture. And expectations and laws regarding EPM make vary significantly from one region to another.

One region with stringent rules about worker privacy is the European Union. The European General Data Protection Regulation (GDPR) states that an employee can expect not to be monitored. When an employer wishes to monitor employees, they must justify why such monitoring is necessary. In applying this law, French regulators have concluded that employers do not have the right to create situations where an employee is under constant surveillance through cameras or keylogging programs.

In contrast, Canadian workers have fewer rights. Charbonneau and Doberstein (2020) describe a popular software program in Canada to monitor at-home workers during COVID. They describe "a video conferencing software that is always on. It takes pictures every few minutes via a front-facing laptop webcam and posts them on a wall so managers can see employees working at their desks" (pp. 780–781). In both the United States and Canada, judges have been asked to rule on whether an employer's actions violate a "reasonable expectation of privacy." Employers can, in some cases, make the argument that monitoring is necessary to ensure that packages are not lost in a warehouse and to guard against theft or breakage (Charbonneau & Doberstein, 2020).

In North America, there does not yet appear to be a clear consensus regarding governments' level of commitment to safeguarding employee privacy as a value.

Issues of Equity and Fairness

There are both positive and negative arguments for employees regarding adopting workplace surveillance mechanisms.

Some disability rights groups have argued that using EPMs violates their rights, including the right to receive reasonable accommodations for a disability. They argue against EPMs, stating that everyone should be judged on the quality of their work and not on how quickly they type or how long they take to perform specific tasks.

However, one might also argue that EPM technology is a positive development since the development of objective, transparent metrics for measuring employee productivity makes it easy for employees to be treated fairly. Using EPM technology that ranks employee productivity could mean that those who might have previously been passed over for promotions and training due to employers' prejudicial or discriminating practices might do well under such a new system. One might even argue that the values of equity, inclusion, and diversity are more important values than those related to privacy and an absence of surveillance.

Emerging Issues

Emerging legal issues around workplace surveillance include using predictive algorithms to aggregate data about current employee behaviors to predict how they might behave in the future. Oliver Roethig, a spokesman for UNI Europa, the European Service Workers Union, is concerned about a situation where, for example, a company might try to predict which workers are most likely to join a union and then act to dismiss those employees (Stupp, 2021). Overreliance on algorithms and EPMs can be detrimental if technology is applied without completely understanding its strengths and weaknesses.

Bibliography

Bresiger, G. (2017, July 29). This is how much time employees spend slacking off. *New York Post*. Retrieved from https://www.news.co.au/finance/work/at-work/this-is-how-much-time-employees-spend-slacking-off/news-story/.

Charbonneau, E., & Doberstein, C. (2020). An empirical assessment of the intrusiveness and reasonableness of emerging work surveillance technologies in the public sector. *Public Administration Review*, *80*(5), 780–791.

Stupp, C. (2021, January 20). Monitoring of employees faces scrutiny in Europe. *Wall Street Journal*. Retrieved from https://www.wsj.com/articles/monitoring-europe-1161138602#.

Willis, T. R. (2020). A fresh start: Surveillance technology and the modern law firm. *Duke Law and Technology Review*, *75*(1), 2328–2396.

Encryption as a Response to Surveillance

One way users have responded to the threat of unauthorized surveillance (by one's government or malicious actors) is by developing robust encryption technologies.

Encryption is defined as "the process of using an algorithm to transform information to make it unreadable for unauthorized users" (Techopedia, n.d.). Encryption is a technology, and it is neutral. It is not necessarily unethical; indeed, we can make many solid arguments for and against using encryption technologies. Some analysts compare it to using an envelope in exchanging written communications. Using an envelope is not unethical, though one could undoubtedly enclose illegal *or* unethical content within an envelope.

On the positive side, encryption creates a more secure system that builds trust and cooperation between users. Like a credit card number, our personal information is encrypted when it travels over the internet as we conduct activities like online shopping. Encryption assures that only authorized users with a key can access and read the data. Thus, encryption preserves people's privacy and provides confidentiality. Encryption also provides **authentication**. Through the use of keys, everyone engaged in a transaction is assured that the people they interact with are who they say they are. Encryption also provides *non-repudiation* by providing proof of which party sent a message and who received it. It also contains time-date stamping, thus proving when the message was created, sent, and received (Roberts, n.d.).

While encryption protects lawful users' information, unlawful users can also use it. **Ransomware** crimes are crimes in which individuals are contacted and told that they have downloaded a piece of malware that will encrypt all of their data unless they pay a fee to the "kidnapper," who then provides the key to decrypt the data. If victims do not pay by a specific date, their data can be destroyed. Many entities – including hospitals – have been the victims of ransomware, with innocent medical patients the victims if their data is lost. Terrorist groups may also use encryption technologies to pass secure communications within a network without observation or monitoring by law enforcement.

In the United States and internationally, encryption norms are still evolving, and nations have adopted vastly different policies regarding the legal level of encryption. In 2016, Great Britain passed the *Investigatory Powers Bill*. This legislation grants authority to the government for bulk collection and lawful hacking. It allows the government to order the removal of electronic protection (encryption) applied by an operator to any communications or data. The legislation was seen as creating the authority for the government to require a "*back door*" that would allow them to bypass any encryption. And in the aftermath of the 2016 terror attacks in France, the government declared a state of emergency, simultaneously passing legislation that would fine technology companies that refused to decrypt the messages for law enforcement. However, within the European Union, the Netherlands and Germany strongly support strong encryption and grant only limited powers to the government to force decryption. In December 2015, China passed an anti-terrorism law that requires telecommunications companies and internet service providers to provide technical interfaces, decryption, and other services to state security. In India, the government has attempted to pass laws banning certain forms of end-to-end encryption, including the WhatsApp application, often used for communications by legitimate and illegitimate actors. However, these proposed laws were withdrawn from consideration in 2015 due to the actions of Indian civil rights groups.

Despite limited support for backdoors and government control over encryption in many nations, in 2016, the United Nations High Commissioner for Human Rights stated that encryption and anonymity help enable human freedom of expression and opinion and the right to privacy. Thus, the United Nations has indicated that they strongly support the widespread availability of encryption technologies to citizens.

Establishing Conditions for Ethical Surveillance

As we have seen thus far in the chapter, one can make ethical arguments for and against surveillance technologies. In the remainder of this chapter, we will consider the virtue ethics, utilitarian, and deontological arguments for and against surveillance. We will also lay out the difference between "good" and "bad" surveillance and the conditions under which surveillance might be considered ethical.

The Deontological Lens

As noted in Chapter 2, the deontological approach to ethics rests on the assumption that treating individuals merely as a means to an end rather than as an end in itself is inappropriate. In this approach, the highest value should be on maintaining the dignity and happiness of individuals.

In considering the ethics of surveillance, a deontologist would ask: How might a computer user feel if they found out that their data was being collected and analyzed? How likely is it to cause the user to feel harm or embarrassment?

In this approach, any long-term payoff that might come from examining users' data without their knowledge or consent would be irrelevant – if it was felt that the users were somehow harmed or made uncomfortable through surveillance. In her work, Landau describes an experiment that researchers carried out related to earlier pancreatic cancer diagnoses:

> Medical and computer science researchers discovered they could anticipate queries concerning the diagnosis of pancreatic cancer based on earlier Bing search queries . . . Pancreatic cancer is tough to treat because it presents late – so this work is intriguing. Could you advise that people go to the doctor based on their queries? Could this lead to an earlier diagnosis? Early treatment?
>
> (Landau, 2016)

She concludes that although some lives might be saved due to the creation of analytics that linked people's symptom queries to specific diseases, the research was unethical. She argues that the research targets would likely feel that their privacy had been invaded if they became aware that their private medical searches were being tracked and analyzed. She notes that no one had volunteered for the study in this situation, nor were they informed that it was being conducted.

In considering who is most often affected by surveillance, we can also reference the work of John Rawls, introduced in Chapter 2. Rawls suggested that the decider should don the "veil of ignorance" in making an ethical decision. He did not know which participant in the scenario he might ultimately be. Then, he should reach a solution where the participants with the lowest status would not be disadvantaged by the decision. If there were any gains from the decision, these gains should again favor the least influential member. In applying Rawls's theory to the question of surveillance, we would identify with the lowest-status individual who is most likely to be the subject of surveillance. How might I think about the utility and need for surveillance if I were an Arab American, a Japanese American whose ancestors had been interned, or an African American man or woman who was already the subject of differential surveillance? What limits might I want to see on the government's ability to carry out surveillance, and what conditions might I place on that surveillance?

Box 5.3 Going Deeper: What Is a Smart City?

What is a smart city? In many parts of the world, city planners and leaders are increasingly making the move towards the "smart city." A smart city is characterized by an elevated level of interconnectivity between systems which collect and use data. Data that is routinely compiled and stored by a city – such as the amount of garbage collected, the traffic patterns occurring on roadways both during normal hours and during rush hour, and the crime rates for different neighborhoods within a city – is increasingly being shared across departments in order to better understand how a city functions as a system. In addition, a smart city may be one that automatically adapts an environment in response to changes which are noted. (For example, lanes on highways might change direction in response to increased traffic, with some roads being used to drive into the city in the mornings and out of the city at the end of the day. And additional police units might be assigned automatically to conduct surveillance or crowd control in response to elevated crime rates or planned events in a neighborhood.) Smart cities may also include elements of participant surveillance, through, for example, allowing citizens to report roads needing repair through using an app.

Many planners champion smart cities, often making an ethical argument. As Smith argues, smart cities are often presented as "virtuous." The cities and their planners may be portrayed as intelligent and responsive. Planners might think of themselves as excellent stewards of citizen resources, which they are now using more efficiently to create cities which are safe, better prepared for disasters, and even more democratically administered through citizen participation. Proponents of smart cities may argue that they are better at solving social and ecological crises – like poverty, crime, congestion, pollution, and waste (Smith, 2020).

But participating in the smart city may require that people give up some degree of privacy – whether they are being asked to show a code on their phones to enter a dog park or highway, or agreeing to have their license plates photographed as they enter and exit the city. Smart cities depend on surveillance, and people may find either that they are compelled to participate in these surveillance activities, or that such activities take place even without citizen awareness and consent.

Smart Cities and Racial Disparities

Another concern expressed by researcher Sara Safransk relates to what she refers to as "the geography of algorithmic violence." She argues that people often see planning algorithms as neutral, simply providing a clear and objective way to decide how best to locate resources like hospitals. However, she argues that both consumers and planners may use devices like maps showing where crime rates are greatest or which schools are failing to make decisions, which lead to greater segregation along racial and class-based lines.

In her work, Safransk describes how the city of Detroit, Michigan (widely known in the United States as an impoverished city in which industries have shut down, businesses have been shuttered, and entire neighborhoods have been foreclosed and abandoned), used data about the city's geography and its residents to limit services like public transportation to some parts of the city. As a result, people who already had few economic prospects had even fewer, once public transportation was no longer available.

Bibliography

Safransk, S. (2020). Geographies of algorithmic violence: Redlining the smart city. *International Journal of Urban and Regional Research*, *44*(2), 200–218.

Smith, G. (2020). The politics of algorithmic governance in the Black Box city. *Big Data and Society*, 7(2). https://doi.org/10.1177/2053951720933989.

Utilitarian Ethics

As noted in Chapter 2, utilitarian ethical frameworks offer flexibility because the right thing to do might vary according to the environment or circumstances. That is, utilitarian ethics allows for the possibility of situational ethics – or the idea that a choice that would seem ethically wrong in one situation might not be so in another.

The most common justification given by the US government for its increasing surveillance practices is that ever since 9/11, the United States has found itself in a highly unusual situation. Since the United States currently has many enemies who wish to harm the United States and its citizens, the argument goes, we should thus consider the sorts of ethics appropriate during wartime. According to this argument, during peacetime, for example, the role of government might be "reined in." Still, during a time of grave danger to American citizens, people might be more willing to give their government more power to watch its citizens and activities. Analysts point to the USA Patriot Act, a piece of legislation passed in the immediate aftermath of September 11, 2001, which ceded extraordinary power to the US government.

One can also consider who surveillance targets. Here, a utilitarian would argue that while citizens should have the right to be free from surveillance, individuals who are suspicious or pose a danger to national security should have their rights curtailed.

In the aftermath of the 2013 Snowden revelations, Britain's former GCHQ chief, David Omand, published an op-ed in *The Guardian*, a British newspaper. He argued that a government could ethically carry out covert surveillance of citizens provided certain conditions were met. (**Covert** here refers to secret surveillance both in the sense that the subject may not know that he is being surveilled and that the conduct of this surveillance may not be a matter of public record within the government itself.) In the op-ed, Omand (2013) made two types of ethical arguments. First, he invoked the concept of duty – stating that the intelligence community was carrying out its duties, which included the imperative that they take measures to keep citizens safe. He then defended the actions of corporations like Google and Facebook that allowed the NSA access to customer communications – again making an argument from duty. He notes that they merely complied with American law and should not be faulted.

In addition to making an argument from duty, Omand (2013) also advanced a consequentialist argument, noting that "most reasonable people would welcome the security that such intelligence can bring to our nations." In this argument, the risk and the costs of failure – to provide security to one's country and citizens – are so significant that depriving citizens of their rights is offset by the higher cost of failing to provide security.

But Michael and Michael (2011) argue that there are many costs associated with surveillance that utilitarian philosophers may have failed to consider. They argue that someone might commit suicide due to being publicly humiliated by the publishing of personally compromising information. They note that a completely transparent society characterized by total surveillance could thus have psychological consequences, citing "increased cases of mental illness – new forms of an obsessive-compulsive disorder and paranoia; a rise in related suicides; decreased levels of trust and the impossibility of a fresh start" (2011, p. 14).

Finally, we can apply the utilitarian paradigm to weigh the costs of surveillance and the opportunity costs to society. What is the likelihood that people might choose not to visit the United States, invest in the United States, or work in the United States due to ubiquitous or differential surveillance? And in establishing ubiquitous surveillance, what opportunity costs would a nation sustain? What other services might a nation have to forgo to pay for establishing an expensive surveillance system? Does it make economic sense to engage in anticipatory surveillance to preempt an expensive terrorist attack, or should those resources be deployed elsewhere?

Box 5.4 Going Deeper: Health Surveillance Ethics

What is health data surveillance, and how does it work? Epidemiology is the science that seeks to understand disease patterns at the aggregate or societal level. Epidemiologists seek to understand how an outbreak of disease might occur and what interventions are most effective in slowing the spread of disease (Duff-Brown, 2020).

Digital epidemiology is a new and growing field in which public health specialists use Big Data analytics to track the spread of disease. Beginning in 2009, with the Google Flu Trends Map, researchers have looked at keywords that people search (like "flu symptoms") to see which regions might be in the early stages of an epidemic of diseases. Today, researchers might also look at social media postings. Other forms of "disease surveillance" include tests that researchers run on waste material samples from waste treatment facilities. Particularly during the COVID-19 pandemic, researchers have identified regions where outbreaks occur by looking at whether disease particles are found in biological waste.

Many epidemiologists argue that people must share their data and participate in data-gathering initiatives rather than opting for personal privacy. And in some situations, like wastewater treatment sampling, people may not even be aware that they are participants in a study, nor have they been asked for their consent (Mello & Wang, 2020).

Collecting Disease Surveillance Data

Particularly with the outbreak of COVID-19 in early 2020, many governments began working with industry to collect the most accurate and up-to-date data about individuals' exposures to this contagion. COVID tracker apps, programs that users can download onto their phones to track their exposure to disease, come in two variants: decentralized and centralized.

Decentralized apps rely on peer-to-peer networks. Contact information is stored on a user's phone and then shared with other users without entering or going through a central repository. Such apps are designed using the principles of privacy by design to preserve user privacy as a central value. Using a decentralized app, a person gets a positive test and informs the app. For every location that she visited in the previous month, her phone had sent a string of text (known as an **ephemeral identifier**) every few minutes to other phones within her proximity. Every so often, phones running the app will connect to a central database containing strings of texts from all users who have reported that they have a positive test. If your phone's stored texts match any of the strings on the central database, you are informed that you have been exposed to COVID (Tech Policy@ Sanford, 2021).

Centralized apps, in contrast, do not use peer-to-peer networking but instead have phone users send location data information to a central server. The central server gives each user a permanent pseudonymous identifier in a centralized app configuration. Each user's ephemeral identifiers are logged and held for a specified period (one month). When a user is diagnosed with COVID, she informs the app, sending a message to the central server. Phone Xchange ephemeral identifiers are then "matched" through the central server, and people who might have been exposed receive a message from the server. A centralized system provides more information to the server, such as where clusters of results occur and how the disease might spread.

In 2020, both Google and Apple announced that they would support only the sale and installation of COVID-tracker apps on their store platforms, which relied on decentralized data storage. Since both Google and Apple devices automatically track user locations, the apps developed rely on location data generated by the phone itself, then collected by the disease surveillance app using an API. The announcements by Apple and Google led the United Kingdom, in particular, to switch over to the development of apps using decentralized data storage to avail themselves of Google support. Australia, Singapore, Italy, and Germany had also begun creating centralized apps, but they also switched to developing decentralized apps to work with these two major providers.

A third option may perhaps be used in the future. Some analysts suggest that the best way to store and share data may be to use blockchain technology. Cybersecurity experts note that the transactions will be more secure and tamperproof by recording people's interactions onto a blockchain, using pseudonymization and ephemeral identifiers. It is significantly less likely that an adversarial actor could somehow wipe out data or alter it if it is not recorded on a centralized server. This technology is thus more resilient and less vulnerable to compromise. Since the blockchain records information in a distributed ledger format, all blockchain participants thus work to create a consensus about whether a proposed record is admissible (using a consensus mechanism to adjudicate the information) (Hasan et al., 2021).

However, algorithmic contact tracing and the blockchain are merely technologies. And while technology can be deployed towards a specific end – like disease surveillance – this is not the only end to which it could be deployed. Suppose you have a program that tracks who is in physical proximity to someone else, often without the user's permission or even knowledge. In that case, it could also be used to accuse someone of having an extramarital affair, visiting a place where sex work is taking place, or engaging in espionage. For this reason, some states (like New Jersey and New York) have begun to pass legislation explicitly safeguarding disease surveillance information and declaring that no other individual, entity, or government agency will have the right to request or have access to this information for any other purpose (Caplan et al., 2020).

Issues of Diversity and Equity

Given the utility of contact tracing apps, is this sufficient reason for states to require their citizens to participate? In discussing the ethics of contact tracing apps, Morley et al. (2020) note that "digital interventions come at a price." For this reason, they propose several limitations which might be imposed upon those who develop health tracker apps. Specifically, they have adopted principles from the European Convention on Human Rights, the International Covenant on Civil and Political Rights, and the UN Syracusa Principles to suggest that contact tracing apps need to be necessary and proportional (that is, there needs to be a sufficient health risk that the harmful risks to privacy are outweighed by the positive benefits of stemming disease outbreak). Such apps also need to be scientifically valid and time-bound (that is, data should be collected only for a prespecified period, and specific rules and regulations should govern how data is stored and how long it can be stored).

Bibliography

Caplan, M., Essey, S., & Bogard, S. (2020, April 20). Location, location, location! – data, privacy and coronavirus. *Gilbert and Tobin Law Blog*. Retrieved May 6, 2022,

from www.gtlaw.com.au/insights/location-location-location-data-privacy-coronavirus.
Duff-Brown, B. (2020, May 11). Ethics and governance of digital epidemiology. *Freeman Spogli Institute for International Studies Blog*. Retrieved May 1, 2022, from https://fsi.stanford.edu/news/ethics-and-governance-digital-epidemiology.
Hasan, H., Salah, K., Jayaraman, R., Yagood, I., Oman, M., & Ellahham, S. (2021). Covid-19 contact tracing using blockchain. *IEEE Access*. Retrieved June 2, 2022, from www.ncbi.nlm.nih.gov/pmc/articles/PMC8545209/.
Mello, M., & Wang, C. J. (2020, May 11). Ethics and governance for digital disease surveillance. *Science*, *368*(1494), 951–954.
Morley, J., Cowls, J., Taddeo, M., & Floridi, L. (2020, June). Ethical guidelines for COVID-19 tracing apps. *Nature*, 29–31.
Tech Policy@ Sanford. (2021, February 21). Comparing centralized and decentralized contact-tracing approaches. *Tech Policy@Sanford*. Retrieved May 2, 2022, from https://sites.sanford.duke.edu/techpolicy/2021/02/21/centralizedvsdecentralized/.

Virtue Ethics

Much of the virtue ethics position on surveillance was presented in Chapter 4 when we examined US Secretary of State Henry Stimson's statement that "gentlemen do not read each other's mail." Here, Stimson argued for the virtue of restraint, a virtue that appears in Aristotle's writing. A virtue ethics position might thus identify specific types of surveillance activities as inappropriate – not because of the harm to the victims that they might generate – but because the conduct of such activities would not be in keeping with the character of a virtuous "spy." For example, the American Civil Liberties Union (n.d.) has described how individuals in authority have misused surveillance technologies – including law enforcement officers who have used surveillance videos to stalk women or threaten and harass ex-spouses, or situations in which an individual might be blackmailed due to video or photo evidence that they have visited a gay bar. All of these represent the use of surveillance for wrong motives.

We can identify elements of the virtue ethics position in the Association of Computing Machinery's own Code of Ethics, particularly in section 1.7, which reads "respect the privacy of others." Regalado (2013) has advanced the argument that ACM members who worked for the National Security Agency in 2013 were thus violating the ACM ethical code since they did not engage in restraint in respecting an individual's privacy. Others disagreed with him, making a practical argument. In response to Regalado, Eugene Spafford, an ACM official and a professor at Purdue University, argued (much as Stimson's critics did in the original scenario) that a greater good may be served if surveillance leads to information that prevents a terrorist outbreak.

What Does Ethical Surveillance Look Like?

In his 2013 article referenced earlier, David Omand, the head of Britain's Signals Intelligence Organization, the GCHQ, proposed six principles that can be used to determine the ethicality of communications intercepts. These principles are loosely based on just war ethical principles, a set of understandings regarding the ethical conduct of war that dates back to the Middle Ages.

Omand proposes first that there must be a sustainable or genuine cause for which communications intercepts are necessary. That is, it is unethical for an agency (or an individual) to merely engage in a "fishing expedition" where they poke around in communications in hopes of finding something illegal to prosecute.

Second, he notes that there must be a motive for conducting surveillance, and that those seeking permission to engage in surveillance should be forthright and honest in declaring that purpose. (They shouldn't lie about their motives or aims.)

Third, he suggests that surveillance methods should be proportionate or limited. Omand argues that those conducting surveillance must be aware of possible harms and determine that the possible benefits outweigh the harm. If possible, they should exercise restraint in deciding who to surveil and under what conditions. Here, Omand utilizes a deontological lens in explaining the doctrine of minimum trespass or minimum force. This idea is borrowed from military ethics. Minimum trespass means the party should strive to do minimum damage against national and individual human rights in an activity. In other words, he should respect the dignity of the individual. Those who collect intelligence write that the collectors should be minimally intrusive and minimally invasive (Jones, 2009, p. 37). In a similar vein, Pfaff (2009) distinguishes between soldiers who are legitimate military targets since they have entered into a conflict-free and are aware of the dangers they are subjecting themselves to, and civilians who are not. In writing about surveillance, he argues that an individual suspected of plotting harm could be a legitimate target of surveillance, but not his family members or others. He notes that "exploiting them may be the most expedient way to get information, but it is not the most moral because none of these groups have knowingly and intentionally entered the 'game' in the way the other groups have." (Pfaff, 2009, p. 83).

In addition, Omand notes that there should be a *rightful authority* – meaning that there should be documentation of surveillance patterns, clearly establishing accountability and a chain of command. This is why ethicists have raised objections to new automated surveillance forms – a lack of rightful authority. The blog *Ars Technica* describes a technology that the US National Security Agency has used since 2010, known as Turbine. The turbine is used to hack millions of accounts a day. It automatically installs malware onto internet-connected devices after targeting individual accounts. Turbine looks for cookies from many different services, including Google, Yahoo, and Twitter, and some Russian cookies from servers like Yandex. Once the "implants" are installed onto a user's system, the NSA and Britain's GCHQ can extract data, monitor communications, and even attack networks (Gallagher, 2014). The program thus creates an ethical distance between any potential harm created (such as an invasion of privacy) and the legally accountable organization (the National Security Agency). In this scenario, one might argue that Turbine was responsible for the invasion of privacy. Since Turbine is not human, it cannot be held morally accountable for the damage it has created.

Similarly, Fogg (2003) describes a situation where a food service company installs a technology to check whether employees wash their hands before returning to work. He argues that an ethical system would notice a breach and prompt the employee to go back and wash his hands. In contrast, a corrupt system would automatically notify the employer, resulting in the employee's termination. In the first scenario, the employee knows he is being monitored and "buys in," cooperating with the software. In the second scenario, the employee may not even know why he was terminated and is unable to place blame on the surveillance system for having harmed him by costing him his job (Fogg, 2003, p. 233).

Finally, Omand argues that there should be a reasonable prospect of success expected when deciding to conduct surveillance and that secret intelligence collection should be a *last resort* when other methods of extracting information, such as diplomacy, have failed. These six points thus include virtue ethics arguments, including the notion of integrity; duty arguments, including the establishment of rightful authority; and consequentialist arguments, such as proportionality and a reasonable prospect of success.

Good vs. Bad Surveillance

So how might a computer programmer asked to engage in surveillance activities think about his actions? We will conclude this chapter by comparing and contrasting ethical and unethical surveillance. As we have seen in this chapter, good surveillance is benevolent. That is, the intent of surveillance matters. It is also controlled and regulated. In surveilling regular citizens – and not those accused of crimes – people should be aware that they are being surveilled. (This is why citizens in Britain, for example, see so many signs on their streets and public transport informing them that they are under surveillance by closed caption television systems, or CCTV.) In some instances, good surveillance might allow people to opt out of surveillance. Finally, good surveillance is not differentiated but is equal opportunity.

In contrast, unethical surveillance might be carried out with bad intentions and might be perceived by those surveilled as intrusive, humiliating, or threatening. It is often asymmetric, with someone in a more significant position of power watching someone with less power to object. Unethical surveillance often does not warn people, even private citizens, that they are under surveillance. It may not provide the ability to opt out of surveillance, and it is often differentiated by social class, gender, or race.

CHAPTER SUMMARY

- The "right to surveillance" belongs to corporations, agencies, and states, while the right to privacy belongs to individuals.
- Ubiquitous computing means that today, more people are watched in more places at more times, and more data is stored and shared. Critics suggest that the "death of privacy" is inevitable and that surveillance and privacy are incompatible.
- People in different cultures may hold different views about acceptable levels of surveillance and privacy. As a result, nations may have different surveillance and citizen rights laws.
- Today, globalized data storage patterns complicate surveillance activities since it is not always clear whose jurisdiction prevails when national laws differ regarding surveillance.

DISCUSSION QUESTIONS

1 **Ethics of Facial Recognition**

Watch Professor Woodrow Hartzog. "The Case Against Facial Recognition." The University of Hamburg. 1-hour video presentation available at https://lecture2go.uni-hamburg.de/l2go/-/get/v/53997 (accessed June 21, 2022).

- Do you agree with him that facial recognition technology should be banned? Which of his arguments do you feel are most compelling, and why?

- What is meant by the phrase "privacy by obscurity"? Do you feel that privacy is vanishing today? Is it inevitable?

2 **In-Class Exercise on Surveillance**
Ask students to go to the Facebook Transparency center and access the following: https://transparency.fb.com/data/government-data-requests/data-types/
This page allows you to see which governments and their agencies have requested user data from Facebook, and the types of data that they have requested.

- Choose a country and look at the types of data the country has requested from Facebook and whether or not Facebook has complied with the requests.
- What are the reasons why a nation might request data from Facebook?
- Are you surprised at how much or how little Facebook complied with these national law enforcement requests for user data? Are you surprised at how many there were? Which countries were MOST likely to request user data from Facebook? What sorts of trends do you notice?

If you are American, visit the following page, which looks at data requested by American organizations related to US national security: https://transparency.fb.com/data/government-data-requests/country/us/ (accessed June 21, 2022).

- What ethical issues do you see raised by these requests and by Facebook's cooperation with these requests?
- In this course, we have discussed competing ethics – what competing ethics do you identify here?

3 **Ethics of Servitization**
In their work, Royakkers and van Est (2015) point to that growing pervasiveness of a phenomenon which they call "servitization" – in which people no longer own goods but instead subscribe to services that give them access to the goods. They warn that if you purchase music access instead of the record album, whoever owns the service can collect information on you and your listening habits.
If "servitization" becomes the norm for your generation, is it inevitable that people will lose control over the data they provide, how it is collected, and how it is stored?
Think about the pros and cons of servitization. What do you as a user receive by having access to a platform that collects all of your music data, your health data or data about your transportation requests (made to a service like Uber) and so on, and what do you give up? Do you feel that the pros outweigh the cons in this situation?

4 **Comparing Centralized and Decentralized Data Storage Protocols**
Compare the ethical pros and cons of decentralized and centralized approaches to data storage for contact tracing apps for disease surveillance purposes.

- What are the pros and cons of each approach?
- What values are being preserved (and whose values)?
- Can you imagine an appropriate situation to compel people to participate in a disease surveillance program using an app? Describe that situation.

RECOMMENDED RESOURCES

Au, Y. (2021). *Surveillance as a service: The European AI-assisted mass surveillance marketplace*. Oxford: Oxford Commission on Artificial Intelligence and Surveillance.

Downey, M. (2019, August 28). Legal ethics and geofencing. *Litigation*. Retrieved from https://www.americanbar.org/groups/litigation/publication/litigation-journal/2018-19/summer/legal-ethics-and-geofencing/.

Fisler, C., Beard, N., & Keegan, B. (2020). *No robots, spiders or scrapers: Legal and ethical regulation of data collection methods in social media terms of service*. Palo Alto, CA: Association for the Advancement of Artificial Intelligence.

YouTube Video (10 minutes): John Arnott. *GeoFencing and proximity marketing*. Retrieved October 19, 2020, from www.youtube.com/watch?v=QXiWaZOM9zc

References

Agre, P. (1994). Surveillance and capture: Two models of privacy. *Information Society, 10*(2), 101–127.

American Civil Liberties Union. (n.d.). *What's wrong with public video surveillance*. Retrieved January 24, 2017, from www.aclu.org/other/whats-public-video-surveillance.

American Library Association. (2008). *Code of ethics*. Retrieved from www.ala.org/tools/ethics.

Amnesty International. (n.d.). *Two years after Snowden: Protecting human n rights in an age of mass surveillance*. Retrieved April 12, 2017, from www.amnestyusa.org/research/reports/two-years-ater-snowden-protecting-human-rigts-in-an-age-of-mass-suveillance.

Au, Y. (2021). *Surveillance as a service: The European AI-assisted mass surveillance marketplace*. Oxford: Oxford Commission on Artificial Intelligence and Good Governance.

Azaria, A. R. (2007). Behavioral analysis of insider threat: Survey and bootstrapped prediction in imbalanced data. *Journal of Latex Class Files, 1*(2), 135–155.

Basu, A. (2015). Why Microsoft's "data trustee" model is a potential game-changer in the privacy war. *The Wire*. Retrieved April 12, 2017, from https://thewire.in/15735/why-microsofts-ata-trustee-model-is-a-potential-game-changer-in-the-privacy-war/.

Bazan, E. (2008). *The Foreign Intelligence Surveillance Act: A brief overview of selected issues*. Washington, DC: Congressional Research Service.

Donahue, K. G. (2012). Access denied: Anticommunism and the public's right to know. In K. E. Donahu (Ed.), *Liberty and justice for all? Rethinking politics in cold war America* (pp. 21–50). Amherst, MA: University of Massachusetts Press.

European Commission. (2016). *Protection of personal data*. Retrieved January 23, 2017, from http://ec.europa.eu/justice/data-protection/.

Fogg, B. (2003). *Persuasive technology: Using computers to change what we think and do*. Boston, MA: Morgan Kaufman Publishers.

Fritcke, E. (2016, October 5). Hacking democracy. *Pacific Standard*. Retrieved April 12, 2017, from https://psmag.com/hacking-deocracy-38d7b2350416#.p0q5h0ydx.

Gallagher, S. (2014, March 12). NSA's automated hacking engine offers hands-free pwning of the world. *Ars Technica*. Retrieved January 8, 2017, from http://arstechnica.com/information-technology/2014/03/nsa-automated.

Gillion, J. (2011). *Overseers of the poor: Surveillance, resistance and the limits of privacy*. Chicago, IL: University of Chicago Press.

Hartzog, W. (2021). *The case against facial recognition*. Hamburg: University of Hamburg.

Hawthorne, N. (2015, May 6). Ten things you need to know about the new EU data protetion regulation. *Computerworld UK*. Retrieved January 23, 2017, from www.computerworlduk.com/security/10-things-yo-should-know/.

Jones, R. (2009). Intelligence ethics. In J. E. Goldman (Ed.), *The ethics of spying* (pp. 13–38). Lanham, MD: Scarecrow press.

Landau, S. (2016). Is it legal? Is it right? The can and should of use. *IEEE Security and Privacy, 14*(5). Retrieved April 12, 2017, from http://ieeexplore.ieee.org/document/7676177/.

Longjohn, M. S.-H. (2009). Learning from state surveillance of childhood obesity. *Health Affairs, 29*(3), 3463–3472.

McMullan, T. (2015, July 23). What does the panopticon mean in the age of digital surveillance. *The Guardian*. Retrieved April 12, 2017, from www.theguardian.com/technology/2015/jul23/panopticon-digital.

Michael, K., & Michael, M. G. (2011). The fallout from emerging technologies: Surveillance, social networks and suicide. *IEEE Technology and Society Magazine*, *30*(3), 13–19.

Morley, J., Cowls, J., Taddeo, M., & Floridi, L. (2020). Ethical guidelines for COVID 19 tracing apps. *Nature*, 29–31.

Neocleos, M. (2003). *Administering civil society: Towards a theory of state power*. London: Macmillan.

Omand, D. (2013, June 11). NSA leaks: How to make surveillance both ethical and effective. *The Guardian*. Retrieved June 1, 2017, from www.theguardian.com/commentisfree/2013/jun/11/make-surveillance-ethical-and-effective.

Perlforth, N. (2014). Hackers using Lingo of Wall St. breach health care companies' email. *New York Times*. Retrieved April 12, 2017, from www.nytimes.com/2014/12/02/technology/hackers-target-biotech-companies.html?r=0.

Pfaff, T. (2009). Bungee jumping off the moral high ground: Ethics of espionage in the modern age. In J. Goldman (Ed.), *Ethics of spying: A reader for the intelligence professional* (pp. 66–104). Lanham, MD: Scarecrow Press.

Regalado, A. (2013, September 13). Cryptographers *have an ethics problem: Mathematicians and computer scientists are involved in enabling wide intrusions on individual privacy*. Retrieved April 12, 2017, from www.technologyreview.com/s/519281.cryptographers-have-an-ethics-problem/.

Royakkers, L. M. M., & van Est, Q. C. (2015). A literature review on new robotics: Automation from love to war. *International Journal of Social Robotics*, 7(5), 549–570. https://doi.org/10.1007/s12369-015-0295-x/

Saran, S. (2016, February 16). Cyber (in)security in India. *Lawfare*. Retrieved April 13, 2017, from www.lawfare.blog.com/cyber-insecurity.

Scheppele, K. L. (1993). It's just not right: The ethics of insider trading. *Law and Contemporary Problems*, *56*(3), 123–174.

Schneier, S. (2006, January 16). Who watches the watchers? *Schneier on Security Blog*. Retrieved December 2, 2016, from www.schneier.com/blog/archives/2006/01/who-watches-the-watchers.

Schwartz, P. (n.d.). *Data protection law and the ethical use of analytics*. Washington, DC: The Center for Informaiton Policy Leadership.

Stoddart, E. (2011). *Theological perspectives on a surveillance society*. New York: Routledge.

Timan, T., & Albrechtslund, A. (2015). Surveillance, self and smartphones: Tracking practices in the nightlife. *Science and Engineering Ethics*, *24*, 853–870.

Turilli, M., & Floridi, L. (2009). The ethics of information transparency. *Ethics of Information Technology*, *11*(1), 105–112.

Vaz, P., & Bruno, F. (2003). Types of self-surveillance: From abnormality to individuals "at risk." *Surveillance and Society*, *1*(3), 272–291.

Waddell, K. (2017, May 2). The internet of things needs a code of ethics. *The Atlantic*. Retrieved from www.theatlantic.com/technology/archive/2017/05/internet-of-things-ethics/524802.

6 The Problem of Intellectual Property

LEARNING OBJECTIVES

At the end of this chapter, students will be able to:

- Describe traditional ethical arguments in favor of the right to own physical property
- Define critical terms in the discussion of IP issues – including fair use, economic right, moral right, piracy, and intellectual property
- Apply the virtue ethics, utilitarian, and deontological lenses in thinking through the ethical issues of intellectual property
- Describe new technologies, including digital watermarking and the creation of NFTs as a way of preserving intellectual property

As we consider the ethical problems of piracy or theft of intellectual property, we can consider five real-life situations involving theft of intellectual property:

- In 2014, several pharmaceutical and medical device companies – including Medtronic, St. Jude, and Boston Scientific – were the subject of hacker infiltration. Investigators believe that the hackers were attempting to steal proprietary information related to medical devices to replicate the technology in China (Lindsay, 2014).
- In January 2016, the International College Board testing organization canceled the administration of the Scholastic Aptitude Test (an exam used for college entrance in the United States) in five nations – China, Macau, Spain, Bahrain, and Kazakhstan – amidst allegations that some students had seen the test questions before the administration of the examination (Schultz, 2016).
- In 2014, several US citizens and foreign nations, most in their early 20s, were prosecuted by the US Department of Justice after they hacked into and stole simulation software used by the US Army to train Apache helicopter pilots. The perpetrators carried out their actions both within the United States and abroad. ("Hackers Charged in Software Theft From US Army, Others," 2014).
- In 2015, four American men pled guilty to stealing more than $100 million worth of intellectual property. Over two years, they hacked into the networks of Microsoft, Epic Games, Zombie Studios, and the Valve Corporation. They stole software, trade secrets, and prereleased copies of games, which they then sold for a profit (Walker, 2015).

What do these cases have in common? As these examples show, the term piracy covers a wide variety of different types of activities. These activities vary in severity, the

DOI:10.4324/9781003248828-8

perpetrator's intent, and the harm they create. Some readers might question whether all actions constitute theft or criminal activity. These examples illustrate the grey areas (Beckedahl & Weitzmann, 2012) created by rapidly developing technologies and less rapidly developing legal regimes in the United States and abroad. They also show the lack of consensus regarding piracy's norms and ethics.

Thus, we can ask: How does the unique environment of the internet facilitate the theft of intellectual property and perhaps even invite it? And should we use the same criteria for thinking about the ethics of IP theft as we do in thinking about the theft of physical property, or different criteria?

In this chapter, we ask several questions: What are the rights of creators to "own" the products they produce online or in the real world? How should cyberspace be governed to allow for the idea of private ownership, and should it be set up in this way? And how should individuals consider their actions in cyberspace – regarding whether and under what conditions they respect the right to private property?

We begin by defining key terms. We then dive into the ethics of ownership and the key differences between cyberspace and real space. We then consider legal and ethical understandings relevant to understanding this problem. We conclude by applying our three models – virtue ethics, utilitarianism, and deontological ethics – to think about cyberspace IP issues.

What Is Intellectual Property?

We begin by considering the broader notion of **intellectual property**. What does it mean to own an idea? Although this sounds like a very modern notion, we can trace the idea of intellectual property back to 500 BC, when chefs in the Greek colony of Sybaris were given a monopoly over their ability to produce a particular culinary dish. Intellectual property is also recognized in British law, going back to the *Statute of Monopolies*, passed in 1624, and the *Statute of Anne*, passed in 1710. The Statute of Monopolies still provides the basis for the American and British patent systems. Nasheri defines a *patent* as "an exclusive right granted for an invention (a product or process that provides a new way of doing something or offers a new technical solution to a problem)." A patent lasts for a specific period, is for a specific geographic area (such as the United States), and requires that the inventor publicly disclose his process or product specifications through filing a patent (Nasheri, 2005, p. 5).

The Statute of Anne established the notion of copyright for literary works. *Copyright* is granted for artistic products and can be given to the creator and passed on to their heirs. Copyright protects artistic works – like novels, plays, photographs, and music – while industrial property laws and regimes protect inventions and industrial designs – through patents and trademarks (Nasheri, 2005).

We see how technological developments made these ideas necessary in considering patents and copyrights. With the invention of the printing press, individuals could make multiple copies of a document. It became necessary to establish the conditions under which a document could be reproduced and shared. Today, with the growth of the internet, it is easier than ever to download and upload files and images. New legal developments establish new understandings of what it means to own ideas.

Anglo-American intellectual property arguments rest on utilitarian theory: US president Thomas Jefferson – himself an inventor and the creator of both the swivel chair and the pedometer – argued that the inventor didn't have a "natural right" to control his output and its use, but that it was a reward granted to him so that society as a whole could progress. In utilitarian theory, society maximizes utility by giving rights to authors as an incentive for progress (Moore, 2005).

As Varelius (2014, p. 299) states, "intellectual property rights protect the financial interests and reputation of creators of intellectual objects – objects like inventions, melodies, concepts, methods and (expressions of) ideas." She notes that the claim that one can "own" an idea rests on the legal idea of ownership of physical objects. Full ownership of material objects includes the right to use, transfer, destroy, and modify them. The owner can also decide who to share or refrain from sharing them with. Varelius distinguishes between *moral rights* and *economic rights*. An economic right is the right to be financially compensated if someone else uses your intellectual property. This is the basis for *licensing agreements*, where, for example, someone who wanted to make and sell a T-shirt or a mug with a cartoon character on it would have to pay a fee to the person who initially drew the image. In addition, the agreement would specify the conditions under which the image could be used and how it could and could not be modified. Nasheri defines a *trademark* as a distinctive name, logo, or sign identifying the source of goods or sources; counterfeiting includes actions to sell a product under a false trademark (Varelius, 2014, p. 5).

A *moral right* is the right to be recognized as the creator of the idea or concept (in other words, not to have your work plagiarized) and the right to control how the object is modified – so that one's reputation is not sullied. (For example, the creator of a cartoon image might object to someone making and selling a pornographic or nude version of that image.)

Within the United States, specific legal doctrines uphold intellectual property rights claims. **Fair use** laws specify the amount of work quoted in a book or document or the percentage of an object (like a song) that can be borrowed without a licensing fee. Copyright and patent procedures allow an artist or creator to claim ownership of a cultural product or idea and copyright it. Patented objects cannot be used without payment of a licensing fee.

However, even though legislation exists within the United States and international agreements like the World Intellectual Property Organization Broadcasting Treaty have been agreed upon, there is still debate about legal and ethical/moral aspects of intellectual property rights. There is no clear consensus among nations regarding the ethics and laws which should govern this area of internet production. Instead, one can identify good and compelling ethical and moral arguments from all perspectives (utilitarian, virtue ethics, and deontological) for and against legal restrictions like copyright.

Box 6.1 Going Deeper: What Is an NFT?

In March 2021, the famed Christie's auction house conducted an auction for a piece of digital artwork known as an NFT. Art lovers bid up the price of this artwork, which eventually sold for $69.3 million. In recent years, several sales of NFTs have exceeded $1 million.

But what is an NFT, and how do they work? Earlier in this chapter, we described the growth of memes, a culture of sharing and modifying existing publicly available digital artwork and assets, and the difficulties of retaining control of one's intellectual property in cyberspace. However, we also stated that the advent of the **blockchain**, a technology for tracking ownership of both tangible and intangible assets, held promise when it came to helping artists maintain control over their intellectual property.

The term "NFT" refers to a non-fungible token, which resembles a license in important ways. Someone who purchases an NFT purchases an intangible good accompanied by a digital contract and chain of custody for that object. The blockchain conveys the contract and creates a chain of custody (or ledger of ownership) for the intellectual property. In this way, the owner of digital artwork or graphic pieces can "prove" that the asset is authentic, which is updated by the artist, whose name is attached. The blockchain allows the owner to demonstrate the asset's provenance, ownership, and authenticity. Each NFT is composed of metadata that makes the asset unique and not interchangeable with other similar assets. (That is, it is non-fungible, the "NF" in the acronym "NFT.") The specifics of the license – or what the owner of the IP can and cannot do with that IP – are contained in the smart contract.

Lewis et al. (2021) describe an NFT as: "a cryptographic tool using a suitable blockchain, most commonly Ethereum, to create a unique, non-fungible digital asset."

Implications

But what are the implications of creating this new type of asset? Patrickson (2021) argues that the ability to register intellectual property on the blockchain will disrupt digital creative industries, particularly regarding innovations by programmers and developers in fields like virtual reality. Blockchain technology can serve as a way to protect non-tangible assets (like lists of customers, data configurations for organizing information, and methodologies for training machine learning models). And, as noted, blockchains are also used to register creative works like a graphic used in a video game or even a tweet!

Some technology issues remain as we envision a world where nearly all digital assets are owned and tracked using NFTs. First, there are currently limitations as to the speed at which blockchain issues can be resolved, and there are also computing power and bandwidth issues related to the storage of NFTs. It can be expensive to store information regarding chains of custody, particularly as they are stored in peer-to-peer configurations.

Economists' Concerns

Finally, economists are beginning to raise concerns about certain aspects of NFTs. (Some think that NFTs are merely a fad and will not come into long-term use.) At the moment, some NFT promoters are excited about their utility as a tool for investment or even a way to make money quickly by getting in on the ground floor of new technology. However, economists worry about the possibility of speculation and a market crash. The problem here is that digital assets are worth only what people are willing to pay for them. A tweet or a digital image has no real-world tangible value. It cannot automatically be traded or exported into another financial vehicle (unlike, for example, changing dollars for Euros at a currency exchange kiosk while on vacation abroad).

And while NFTs are registered on the blockchain, registering an NFT does not automatically mean that no one will ever copy the digital image you own. However, NFT proponents argue that there is a difference between having the original Mona Lisa (which currently hangs in the Paris Art Museum, The Louvre) and having a picture of the Mona Lisa reproduced on a coffee cup. They argue that one is priceless by its authenticity, originality, and uniqueness, while the other is not. The first is, they argue, an investment and an asset, which will increase in value.

A final consideration is the possibility that creating an NFT will make it easier for artists and creators to profit from their work by selling it to others or retaining greater control over their intellectual property.

Bibliography

Lewis, L., Owen, J., Fraser, H., & Dighe, R. (2021, June 29). Non-fungible tokens and copyright law. *Digital Business Law*. Retrieved June 22, 2022, from https://digitalbusiness.law/2021/06/non-fungible-tokens-nfts-and-copyright-law/.

Patrickson, B. (2021). What do blockhain technologies imply for digital creative industries? *Creative Innovation Management*, *30*(3), 585–595. Retrieved June 22, 2022, from https://doi.org/10.1111/caim.12456.

What Is Piracy?

Piracy refers to practices by which individuals upload, download, share, transmit, or distribute copyrighted audio or visual information files. Piracy refers to the transmittal of digital information. Individuals engage in piracy whether they are uploading or downloading information to unauthorized websites, using a program to share these materials from one person to another, or making an audio file from a video that might be online. The *piracy rate* is defined as "the number of pirated software units divided by the total number of units put into use," or the percentage of software acquired illegally. Every nation is estimated to have at least a 20 percent piracy rate, with two nations having a piracy rate of 90 percent (Akman & Mishra, 2009).

Piracy costs in the United States are estimated at $12.5 billion annually. The Motion Picture Association puts its economic losses at $3 billion per year, noting that piracy means losing over 70,000 jobs in the recording industry. States and localities also incur losses since those purchasing bootlegged copies of videos or music do not pay sales tax, which goes back into the community (Moustis & Root, 2016). The Business Software Alliance notes that almost 20 percent of US business software is unlicensed, leading to a monetary loss of $10 billion annually. This organization notes that "when you purchase software, you do not become copyright owners." They argue that a user purchases rights to use a copy of the software, but not to distribute it without authorization. The BSA does not distinguish between large-scale redistribution and sharing copies with friends. Both are regarded as unlawful or unethical (Business Software Alliance, n.d.).

While content producers agree on what constitutes unauthorized use of their products – or piracy – not everyone agrees regarding the definition of piracy. The only international definition is included in the *United Nations Agreement on Trade-Related Aspects of Intellectual Property Rights* (the so-called *TRIPS Agreement*). The agreement notes that:

> Pirated copyright goods shall mean any goods which are copies made without the consent of the right holder or person duly authorized by the right holder in the country of production and which are made directly or indirectly from an article where the making of that copy would have constituted an infringement of a copyright or a related right under the law of the country of importation (art 51, n 14).
>
> (from UNESCO.org)

Thus, piracy is regarded as a criminal activity that violates specific copyright laws in the country of origin and internationally. But why is it unethical?

The Ethics of Property Ownership

In the West, our thinking about what it means to own something and what gives someone an ethical claim to ownership derives from the thinking of John Locke, an English political philosopher from the seventeenth century. However, as Tavani (2005) notes, analysts today disagree about whether Locke's thinking is still relevant in a world of intangible cyber assets. When Locke (1632–1704) wrote his theory of property, he was concerned with the dispersion of *tangible assets* (things that can be felt, seen, and touched). He asked: What are conditions under which someone can claim to own something? And how do we distinguish between those owned in common and those that someone might have a right to own? Finally, he asked whether there were any ownership limits. In other words, could someone rightfully (legally and ethically) claim to own all of something if doing so meant that others might be deprived of the ability to own that thing?

In his *Second Treatise*, Locke argues that you earn the right of ownership through taking an object (like farmland) and "mixing (your) labor with it." A village might hold land in common, but the farmer who farms the land can claim to own the land and the vegetables he grows. Here, Locke argues that because we own our bodies – and have the right to do so – if we mix our bodily labor with that object, we can also claim ownership (Himma, 2013).

However, as Tavani points out, Locke didn't believe that this claim to ownership was unlimited. He clearly stated that you couldn't take all of something (like fruit picked from a fruit tree) if you didn't intend to use it but instead would waste it. He cautioned against hoarding something up, which he viewed as unethical. He also cautioned against taking all of something, believing you should leave "enough and as good" for others (Tavani, 2005).

Applying Property Ethics to Cyberspace

While many scholars begin with Locke discussing intellectual property, others argue that intellectual property is so different from tangible assets that the argument doesn't fit. Here they identify two problems: Some scholars ask if using one's mind to write code or invent an app is the same as "mixing one's labor" with something physical. Tavani (2005) asks whether having an idea come to you while sitting on the beach is the same as toiling in a field all day. He describes this as the problem of "the indeterminacy of labor" (p. 89). Next, some scholars object to the fact that there is no actual tangible "thing" with which a creator or inventor mixes his labor. For example, in composing a symphony, what exactly is the creator using as the raw materials? Hardy et al. (2013) describe the differences between tangible and nontangible objects and how these affect people's perceptions of ownership and property. They describe the relationship between physical property and physical boundaries, noting that you steal something when you remove it from someone else's store or home. There is no similar "signal" in cyberspace to connote that you are stealing something. They also note that when a physical object is taken, there is less of it for everyone else, whereas, in cyberspace, a text, a snippet of code, or a graphic object is not taken but cloned. The "thief" is merely making a copy of something, not removing it from its original location so no one else can use it there.

Others, however, find Locke useful for analyzing IP theft since it considers problems like scarcity and hoarding. These scholars suggest that regulations like copyright restrictions and paywalls, in particular, are the intangible equivalent of hoarding – since they can serve to close people out of particular areas of the internet. They describe the internet as an *information commons*, the intangible equivalent of a patch of shared farmland that abuts a village. Just as one man's decision to take all the farmland might leave others worse off and perhaps even kill them through starvation, these scholars argue that a firm's decision to wall off large parts of the internet through charging access fees could potentially harm others, leaving them out of the information revolution and obstructing their abilities to engage in commerce or learning. (For example, if we conceive of the internet as a conduit for information that would make us all better and more informed citizens, then a system where all of the top newspapers charged a high access fee might mean that, in reality, the only free news available was of poor quality, and citizens were not able to inform themselves using online resources effectively.) In its most extreme formulation, the information commons argument suggests that any attempt to regulate copyright and protect intellectual property on the internet is unethical (Himma, 2011).

Here, Tavani (2005) suggests that we apply Locke to cyberspace by asking two questions:

> Does a particular law or policy diminish the information commons by unfairly fencing off intellectual objects? And are ordinary individuals worse off due to that law or policy when they can no longer access previously available information?
>
> (p. 92)

Here some analysts suggest that we all have an ethical responsibility to preserve the information commons through sharing objects and not recognizing copyright. In recent years, we have seen this stance reflected by a movement in the medical community which seeks to make medical information accessible to everyone who wants to read it rather than to allow it to be placed beyond paywalls where only those with access to subscriptions to medical journals or a medical library can read it (Maisel, 2014). Here, activists have suggested that since people pay taxes used in part to fund medical research through groups like the National Institutes of Health, all people should have a right to read any research that might result from these grants. In the most well-known case, libertarian computer activist Aaron Swartz bulk-downloaded the JSTOR scholarly database archives using a guest account on the Massachusetts Institute of Technology's computer network. Swartz, a co-founder of the online discussion board Reddit, was threatened with prosecution for his 2010 actions. He has been described as a "martyr" of the open access movement, as he committed suicide in January 2013 at age 26. JSTOR eventually made 4.5 million articles from its archives available for free to the public. Subsequently, many institutions – like the National Institutes of Health, the Massachusetts Institute of Technology, and Harvard University – have implemented open access mandates – requiring research the institution helps fund to be made openly available through open access journals or in an institutional repository (Hockenberry, 2013, p. 7). Swartz's actions in fighting for open access have been described as a form of civil disobedience. In his Guerilla Open Access Manifesto, he referenced people's moral obligation to oppose unjust laws. He saw the *Stop Online Piracy Act* (*SOPA*) as an unjust law.

Box 6.2 Critical Issues: COVID and Intellectual Property

Is health a global human right, and does this mean everyone should have access to vaccines and medicines, regardless of where they live and how much money they have? And if so, what is the best way to assure that everyone has access to medical care, particularly in a global pandemic?

With the outbreak of the COVID-19 virus beginning in winter 2020, many ethics questions became salient, particularly in the field of intellectual property. Historically, pharmaceutical research has been carried out, in North America and Europe, by for-profit corporations, sometimes working in cooperation with governments and universities. New drugs and therapies are patented and then sold for a profit. Profit serves as an incentive for researchers and corporations to spend the money and the time developing these products.

But what happens when people need access to new drugs and vaccines as quickly as possible and the existing companies cannot manufacture enough quantities of these needed products on a global scale? This was the debate that occurred at the World Health Organization in 2021.

WHO and the Technology Access Pool

The World Health Organization undertakes many programs to ensure global health, including participating in disease surveillance and monitoring medical care quality. But the WHO also has a series of voluntary programs through which corporations and nations can share information about vaccines and medicines with colleagues in the developing world. During the COVID-19 pandemic, the WHO developed new programs, including the COVID-19 Tools Accelerator and the WHO COVID-19 Technology Access pool. In addition, it urged companies and nations to "pool" their information about COVID-19, rather than keeping it themselves as a matter of intellectual property, by using the Medicines patent Pool and Tech Access Partnerships (Sparsh, 2021).

As Sparsh noted, the World Health Organization undertook many new initiatives, such as fast-tracking the procedures for approving the use of experimental vaccines – all to increase global citizen access to vaccines and health care. In doing so, the WHO prioritized the values of equity and access, deciding that the risks to quality were less significant than the need to provide as many people as possible with vaccinations as quickly as possible. To make vaccines available as quickly as possible, she argues, it was decided to share information about how to manufacture vaccines and medicines with colleagues in the developing world – rather than waiting until the five major vaccine manufacturers had sufficient capacity to manufacture a global supply themselves. For this reason, the pharmaceutical company Moderna, which had patented its vaccine, declined to enforce its patent.

Increasing Global Supply

However, some critics of the decision to stop enforcing IP rights for vaccines argued that shortages of vaccines were not because manufacturers didn't have access to information. Instead, they argued, making technical information available was akin to providing people with a recipe – while failing to acknowledge that they may still lack a kitchen, skilled cooks, or the right ingredients to make a dish (Boldrin & Levine, 2021). These critics also worried that knowing that one's intellectual property should be shared this way might be a disincentive for people to conduct such research.

Finally, they were concerned about a decision to waive concerns about quality, efficacy, and safety to achieve the values of equity, efficiency, and speed.

Others noted that there are historical precedents for nations deciding to create a patented technology pool when confronting global health threats like HIV-AIDS, hepatitis C, and tuberculosis. They note that when technologies are pooled, the drugs can often be produced at a lower cost, allowing more people worldwide access to them (de Villemeur et al., 2021).

Many experts credit the decision to pool information about the COVID-19 virus and the COVID-19 vaccine manufacturing process with the ability to develop world nations to make vaccines and medicines available to those sickened by the disease. While intellectual property rights still matter in today's world, sometimes in unprecedented times, other values may prevail, and creative solutions may be found so nations can work efficiently and cooperatively together.

Bibliography

Boldrin, M., & Levine, D. (2021). Reforming patent law: The case of COVID-19. *Cato Journal*, 773–784.

de Villemeur, E., DeQuiedt, V., & Versavael, B. (2021, March 25). Pool patents to get covid-19 vaccines and drugs to all. *Nature*, 529.

Sparsh, S. (2021). The debate around the access to vaccine and licensing amidst second wave of covid-19 in India. *The Journal of World Intellectual Property*. Retrieved June 3, 2022, from https://pubmed.ncbi.nlm.nih.gov/34548844/.

However, others argue that the analogy of information commons/physical commons is not apt. Himma (2011, p. 7) notes that while the farmland available to a village is preexisting (provided by the universe or a god), the resources we find on the internet are created by content providers. The news is available because someone wrote it and gathered it, and that person needs to be compensated for the products of their labor.

Box 6.3 Critical Issues: What Can a Robot Own?

An exciting new legal question relates to inventorship, ownership, and property rights in situations where the creator of a product is artificial intelligence.

Today, AI-enabled programs may write music indistinguishable from that written by humans. Science fiction is becoming a reality, as humans may soon read novels written by artificial intelligence. Artificial intelligence may design sculptures, monuments, and buildings. At the same time, companies may desire to patent algorithms, processes, and programs for dealing with data that have been developed through artificial intelligence.

But who should own these inventions and profit from them? Who should have creative control over how such products are used? (In recent years, we have seen legal disputes regarding whether, for example, the descendants of a composer can license his music to be used as a jingle for a fast-food company. Could an AI specify how music it composed couldn't be used?)

Here an essential related legal issue is that of liability and responsibility. For example, if a driver is involved in a fatal accident, he might seek medical damages from the individual who designed the automobile's operating system. But what if the designer was not a human but an AI? Military ethicists have begun to ask similar questions about who might be legally responsible when an AI-enabled weapon has harmed a civilian. Currently, military regulations in most nations designate that there must be a "human in the loop" when AI is used in warfighting.

Robots Producing Goods and Inventions

Locke's definition of ownership states that one marries one's labor with an object (for example, by forming a field). The addition of one's labor confers ownership of that property. Here clearly, the AI is performing the labor. And, as Nezami (2017) point out, legal frameworks have recognized the possibility of the "legal person" existing, to whom ownership and property rights might be conferred even if the "legal person" is not a human. For example, a university corporation might be the owner of the patent for work developed by university staff using university resources. Therefore, an AI could be designated as a "legal person." And artificial intelligence could participate in relationships with the state by, for example, making profits through an invention and then paying taxes on these profits. In this way, AI could be seen as a legal subject.

Others, however, argue that since AIs don't have their own free will, they cannot participate in legal relationships. (Similar arguments exist regarding whether an AI could someday receive citizenship, vote, or run for office.)

Here, some analysts distinguish between different types of artificial intelligence and the capacity each has. Narrow artificial intelligence refers to the capacity to perform one or more tasks humans can perform (such as alphabetizing a list of patients at a hospital). Strong or general AI refers to the capacity to solve all tasks performed by humans. In contrast, artificial superintelligence refers to robots much more intelligent than human intelligence in every field. In the future, one can conceptualize an artificial superintelligence with the capacity to invent new medicines and vaccines, develop new medical tests and types of imaging, or even cure cancer. Who should own the products of these scientific advances?

When Technology Is Used for Bad and Not Good

And how do we safeguard and ensure that AI-enabled technology, including inventions, are used for good and not evil? How might we assure that the AI produces a vaccine, for example, and not a biological weapon? An early example of computer technology being used for evil purposes is the deployment of IBM's punchcard technology by the Nazi regime in Germany. Nazi administrators used this technology to track prisoners and send them to workcamps, and to organize census data to identify members of groups like Jews, Roma, and sexual minorities. This data was then deployed to engage in the genocide of these groups. In legal proceedings after World War II, the US Department of Justice investigated whether to hold IBM or its officials legally liable for this technology's actions. At that point, the assumption was that the company was the patent holder for the technology. Therefore the humans who held the patents were the logical legal subjects in these proceedings.

Today, however, some ethicists argue that the robot or AI might be conceptualized as similar to an enslaved person who works for an owner. They argue that in Roman law, from which much of our present-day law descends, an enslaver could legally own the products of people's labor he had enslaved. The slaveholder was responsible for maintaining enslaved individuals and was also liable for their actions. The AI would thus be conceptualized as a second-class citizen, like an enslaved person or a child.

Finally, a compelling legal dilemma concerns people who might be augmented with technology in the future. If a person presents as a hybrid since he is married together with technology through physical implants, who might be said to have invented a product in this instance? Should the patent belong to the person or the AI?

These issues are far from resolved, and technology developments may happen much faster than the law can keep up. Nonetheless, it is essential to be aware of these developments and to begin to raise these intriguing questions.

Bibliography

Nekit, K., Tokareva, V., & Zubar, V. (2020). Artificial intelligence as a potential subject of property and intellectual property relations. *Revista de Derecho*, 231–250.

Applying the Lenses

What do the three ethical lenses introduced in Chapter 2 – virtue ethics, utilitarian ethics, and deontological ethics – say about intellectual property?

Virtue Ethics

As we saw in Chapter 2, virtue ethics prioritizes developing moral or ethical character. It suggests that the most ethical decision aligns with the decider's values, allowing them to develop these values. In considering intellectual property, we might reference the value of "care for others" and argue that intellectual property violations are unethical because they expose others to risk. For example, former US Deputy Undersecretary of Commerce for Intellectual Property Stephen Pinkos has stated that 10 percent of all medicine sold worldwide is counterfeit. He argues that states and individuals thus have a duty to oppose counterfeiters since their actions can endanger others (Pinkos, 2005).

Another virtue to consider is integrity – or the idea that your actions and values should be consistent across environments. It would thus require treating all acts of IP theft as equally wrong – not distinguishing between "borrowing" a photo and installing bootleg software on all of one's office computers. In addition, virtue ethics forces us to consider the related ethical issue of collaboration, or "aiding and abetting," in the commission of wrongdoing. Urban (2015) argues that for an event like pirating software from a legitimate vendor to occur, multiple individuals and groups must cooperate. He suggests that everyone who cooperates – through making a website available as a space for illegal transactions, passing along information, or purchasing a good that they suspect is stolen – is both legally and morally involved and therefore responsible for the consequences which ensue, regardless of their small or large part in it.

Box 6.4 Tech Talk: Digital Watermarking of Intellectual Property

Anyone who has ever contemplated making an invention has undoubtedly considered what needs to be done to ensure that they receive credit for their work and that they are the recipients of any financial remunerations associated with it. No one wants to have their hard work stolen by others who profit from it. Therefore, most nations have some form of the national patent office where inventors can register their inventions to receive credit and compensation for their work. Until recently, however, inventors usually received patents linked to creating a physical object – like a machine part, a toy, or a consumer product.

However, today an "invention" might refer to a computer program that someone has authored, as well as a new technique for teaching an AI to carry out a process like sorting or predicting information or storing information. Such intellectual products may be expensive to produce and usually require a significant investment of time and resources. Unfortunately, it is easy for an opponent or adversary to clone or copy information or even copy processes or programs in today's world. We can all think of situations where someone uses an unattributed copy of someone else's work. In many instances, the original creator may not even know that others are deploying their work for new purposes.

For this reason, technology specialists and legal scholars are increasingly interested in techniques for developing so-called digital watermarks so that the creators of intangible products (like programs, text-based information, and processes) can claim what is theirs and prevent others from using it without permission and the granting of credit and compensation.

In 2018, the European Patent Office began reviewing guidelines regarding how creators could patent AI and machine learning techniques, including techniques for creating DNNs or Deep Neutral Networks (Li et al., 2019).

What Is a Watermark?

For this reason, technology scholars have developed methods of protecting people's work.

Steganography techniques allow producers to hide new information within original content. *Cryptography techniques* can be deployed to make a document unreadable or programming capable of being deployed by applying techniques of substitution or permutation.

Watermarking refers to a process where the creator applies techniques allowing them to mark original content as belonging to them. These techniques frequently involve the insertion of marks, images, or techniques to manipulate content in ways that are not visible to readers but can be read by the original creator, who can then prove her connection to the material. For example, watermarking may involve embedding a unique code in text contents associated with the author's secret key.

In the words of Rizzo et al. (2019), watermarking "binds the artifact to the original owner." Various digital watermarking methods have been proposed for identifying and attributing ownership of deep neural networks.

Different kinds of watermarks exist. *Zero-watermarking* does not add content to the original document but instead puts information about the content into an Intellectual Property Right (IPR) database, which creators can access to prove ownership.

Image-based watermarking has been used since the 1990s. This technique may involve shifting text, changing the luminance of pixels, or using other techniques to make slight alterations that only the creator might be aware of.

Syntactic methods focus on altering content elements such as sentence structure. In contrast, *semantic methods* rest on making slight changes to the meanings of words in text blocks. All of these watermarking techniques can also be combined.

Structural methods do not alter text content but make slight alterations to a text structure, such as adding whitespaces.

Blind watermarks cannot be seen with the naked eye but are relatively invisible and imperceptible. However, blind watermarks can be identified using sophisticated techniques, allowing the author to claim her work.

Challenges and Implications

In applying watermarks, some challenges include the best ways of embedding a watermark in situations where only a tiny part of the text or program is used.

Using watermarks to prove and claim one's intellectual property is a technique that has vast national security implications. Since nation-states' adversaries often carry out commercial espionage or intellectual property theft to secure a strategic advantage in both peacetime and wartime, it is essential to develop techniques that allow authors to maintain ownership and control over their digital intellectual property, to assure that it is used and deployed in line with the best interests of their nation.

Bibliography

Li, Z., Zhang, Y., Hu, C., & Guo, S. (2019). How to prove your model belongs to you: A blind-watermark-based framework to protect intellectual property of DNN. In *ACSAC'19: Proceedings of the 35th annual computer security applications conference* (pp. 126–137). Retrieved June 3, 2022, from https://dl.acm.org/doi/10.1145/3359789.3359801.).

Rizzo, S., Flavio, B., & Montesi, D. (2019). Fine-grain watermarking for intellectual property protection. *EURASIP Journal of Information Security, 10*(1). Retrieved June 3, 2022, from https://doi.org/10.1186/s13635-019-0094-2.

However, if one adopts a nonwestern mindset, one can also make a virtue ethics argument in favor of violating intellectual property. Chien (2014, p. 124) describes an Asian mindset in which objects might belong to a family or a community rather than an individual; where one might copy a sage or scholar as a sign of respect, not wanting to risk making a mistake by paraphrasing his thoughts; and where an author might be viewed as having a responsibility to share his knowledge with society without requiring citation or acknowledgment.

Thus, in cultures unfamiliar with western notions of intellectual property, it may be challenging to establish a consensus about the ethics of such an action. Indeed, Ocko (2013) argues that one could frame a virtue ethics argument in favor of "sharing" information within the Chinese context. He notes that in a culture that values collaboration, what we view as an intellectual property violation might be seen as an admirable gesture rather than something to be eschewed. In addition, we can identify divides along economic lines and cultural lines. In developing countries, the "borrowing" of unlicensed software or downloading content like journal articles or video entertainment may be more tolerated since individuals and corporations often cannot afford to pay the fees associated with acquiring these goods legally. Here, they may argue that it is a matter of justice and equity. It is not fair that a scientist should miss out on reading about the latest biology or chemistry advances just because his university is poor. Thus, he will perhaps ask a colleague in a wealthier university to share a copy of an article with him. The wealthier colleague, in turn, may view his act as one of benevolence and sharing rather than a copyright violation (Ocko, 2013).

Mancic (2010) makes a similar argument, asking us to consider the case of the virtuous or altruistic pirate. He argues that often "cyberpirates do much better to the world than copyright or patent holders," arguing that if anyone is immoral, it's the people who refuse to share lifesaving information. The author, a Serbian citizen, writes that many people today cannot buy a book they need to improve their surgery skills or purchase some software that might help a company improve its business and employ more people. This is where "pirates" step in. In his work, he distinguishes between three kinds of pirates: those whose motives are pure to make a profit through selling counterfeit goods or information; those who want to disrupt the system for political or criminal ends, such as terrorism; and those who "pirate" out of compassion, such as wanting to make medical information or proprietary information used to manufacture lifesaving devices available to all who need this information or device. Here, he describes their activity as altruistic, since they simply wish to help the less fortunate. His argument thus forces us to consider the matter of intent.

The Utilitarian Perspective

A utilitarian framework asks whether the individual gains through piracy outweigh the damage to the collective. Crimes like intellectual property theft and plagiarism of one's ideas disincentivize employees and corporations whose job it may be to make discoveries and knowledge. Often many firms will need to cooperate to bring a new product to market. However, if the team members don't trust one another not to steal or misuse their contributions, they might decide not to work together, because the risks of loss are too high. There is less guarantee that you will receive the payoff you expect, including the profits from research and development. In nations that do not have a strong tradition of upholding intellectual property rights, firms worry about whether they will get a patent recognizing their right to own a particular idea, the scope of the patent they may receive, how long the patent might be good for, and how much the new knowledge is worth and how it might potentially be used. Czarnitzki et al. (2015, p. 185) argue that stable intellectual property regimes help all players create and enjoy trust, enabling them to cooperate and share information.

Furthermore, intellectual property theft costs consumers who purchase the product more since prices may be raised to cover losses. We can think of those who profit from a good without paying for it as "free riders." In viewing a movie without helping to fund its production through purchasing a ticket, or in using software without helping to fund research and development through purchase or subscription, nonpaying users harm not only the producers of the good but also those other individuals or groups who paid for the good. Theft raises the price of a good for legitimate users and may hamper research and development efforts, which could create a better product in the future.

In addition, cyber-attacks aimed at securing access to intellectual property – including financial information and trade secrets – can be part of a larger strategy of acts short of warfare carried out by an adversary – either a nonstate actor or a nation-state. Thus, intellectual property violations need to be taken seriously since it is not always apparent whether they are separate acts or part of such a larger strategy. In this case, the practical argument is that "an ounce of prevention is worth a pound of cure." In other words, nations should preemptively combat intellectual privacy violations before becoming more dangerous and challenging (Shackelford, 2016).

However, one can also make a practical argument against copyright protection. In this view, if we view creativity as a deterministic process, in which new ideas are constantly created and old ones are constantly modified – to produce the most social goods for everyone in society – then notions like copyright and fair use may be seen as an unfair barrier or impediment to this process. They keep new content, which benefits all in society, from being created. Any damages to the ownership and reputation of the original creator are therefore seen as a cost worth absorbing.

Indeed, many practitioners today are highly supportive of practices in which **open-source code** is shared among programmers without charge or even needing permission. Even the US government has frequently utilized open-source code in configuring government agencies' websites and programs used by the federal government. Here, federal government procurement officials note that it is unreasonably expensive and wasteful to create new code snippets from scratch (to "reinvent the wheel") when existing programs already solve technological problems, including those faced by the federal government. They also note that the government acquisition process for hiring contractors and writing new programs can be bureaucratic, slow, and unwieldy. In addition, ideally, all federal government agencies and programs would be interoperable and run together and communicate. This is more likely to happen with open-source code than with practices in which all agencies write their code. Thus, many analysts see the practice of utilizing open-source code as efficient and logical. In 2016, the US government implemented regulations requiring that custom code written for one federal agency be made available for sharing across government agencies to reduce redundancies and waste (Bohannon, 2011).

The Deontological Lens

As noted in Chapter 2, deontologists search for a categorical imperative or a rule which could be applied in all situations. Spinello (2003, pp. 17–19) utilizes the categorical imperative to think about piracy in his work. He argues that if you engage in piracy, you must be okay with allowing others to borrow the output of your creative work without compensating you for it.

A deontological argument favoring intellectual property laws would focus on those who would be harmed if everyone felt free to commit intellectual property theft. Here one can build on the Hegelian argument that personal or private property is a means of personal freedom and self-expression (Himma, 2011). Hegel felt that people should have a right to own things since this allowed them to experience meaningful lives. Private property was, thus, a way to experience human flourishing. Thus, one can argue that individuals whose work is taken from them and used or misused by others will have been robbed of something valuable, which decreases their quality of life.

But like utilitarian and virtue ethics lenses, the deontological lens can also be used in two ways – both to condemn and defend piracy. In Chapter 2, we looked at John Rawls's ideas about justice and equity. He recommended making decisions through the veil of ignorance, in which one did not know what one's role or position in a given scenario was. If we adopt the veil of ignorance, then we might view ourselves not as the software developer, but as someone in the developing world who might never be able to "legally" afford a subscription to an online medical journal or access to a library database, but who nonetheless would profit significantly from being granted such access. In this situation, one might argue that the most just or equitable action would be to look the other way if one suspected that unauthorized access was taking place, rather than seeking to rigorously root out and punish all cases of unauthorized access or use (Himma, 2011).

To create an equitable and just internet, these analysts argue that everyone should be able to access the same information online without worrying about paywalls or fees, since it makes little sense to talk about a right to information in cyberspace if many users cannot afford to access that information. Here one can argue that everyone has the right to inform themselves from credible sources, particularly in a democratic society. Those who seek to regulate cyberspace should ensure that everyone has access to more information rather than less.

Adam Moore also calls our attention to inequities – in this case, the privilege that more prominent actors like states and corporations have in cyberspace compared to individual users and small groups. He argues that states and corporations use copyright and intellectual property laws to control who can access and consume information through erecting paywalls and firewalls. For this reason, Libertarians (those who believe that government and regulation should play only a limited role in public life) often refute conventional notions of intellectual ownership – seeing these claims as a threat to individual liberty and rights (Varelius, 2014, p. 300).

Today, much online piracy takes place through peer-to-peer networks. Users can exchange files directly with one another through directly accessing each other's hard drives, sometimes mediated through a server that may be located in a country with weak copyright restrictions. While corporations argue that such technologies allow for large-scale intellectual property theft, some activists claim that they provide an essential service by allowing people to subvert the government and corporate authority by connecting them directly. Thus, they frame their activities as rooted in civil disobedience, in which activists organize to oppose laws they view as unjust and inequitable.

In his work, Mancic (2010) says you cannot categorically say that piracy is always wrong or admirable. Instead, he argues for case-specific ethics – arguing that sharing information about how to make a bomb would be wrong, but that sharing information that heart surgeons in developing countries can use to save lives is not. He argues that the violation depends not on the action but the nature of the content being pirated.

Arguments Against Intellectual Property Protection

Despite the compelling arguments made here – regarding the rights of authors, producers, and other consumers – today, there is no solid international normative and ethical commitment to preserving individual intellectual property rights. But why has it been so difficult to establish an international consensus regarding property?

Those who oppose copyright protection make several arguments in addition to those who have explored here using the three lenses. Many of these arguments cross disciplines, borrowing from moral, literary, and political theories. In some instances, critics of copyright restrictions on the internet raise the question of what it means to be an "author" today. Here they borrow from literary theory and the work of Michel Foucault. In the 1970s, Foucault and others made a splash in literary criticism circles by suggesting that a piece of literature may have no exact, foundational meaning. They argued that the story might not be "about" only one thing but would be interpreted differently in different cultures, languages, and eras. As a result, Foucault argued that one could not rightly speak of only one "author" of a particular work, since the act of reading involved labor on the part of both the writer and the reader, and that finished product, the making of meaning, was a collaborative effort by both participants (Himma, 2011).

Today, the authorship argument is used to defend people's rights to reuse and recycle content they find on the internet, modifying it to make memes and mash-ups. Chanbonpin calls our attention to hip-hop as a unique cultural artifact that draws upon improvisation tradition. He argues that this music style, in particular, is not well suited to be judged by dominant western intellectual tradition and understandings of property (Chanbonpin, 2012, p. 620). These arguments suggest that a new environment (the internet) has created new art forms (like memes and mash-ups) which cannot be treated either ethically or legally in the same way that traditional forms of property and property ownership have been treated.

Two particular cultural products illustrate this point: the meme and the mash-up. In his book *The Selfish Gene*, Richard Dawkins first used the term *meme* in 1976. The meme is a unit of cultural material that spreads virally, passed from person to person via electronic means. It includes phrases, audio and video files, photos, and images (Whatis.com). As it spreads, it may be modified or changed. For example, one can find pictures online of former President George W. Bush falling off a Segway riding vehicle in 2003. The image was passed around the internet, with later users adding additional photos of individuals riding the vehicle, including a chimpanzee and Barbara Bush.

The term *mash-up* is more commonly applied to audio productions. A rap artist might "sample" a track, using snippets and pieces of recognizable older songs as a background or rhythmic accompaniment to a rap. These snippets might be altered from the original by altering the tempo, key signature, or instrumentation. As Rosen, writing on the *Yale Law and Technology blog*, explains:

> Mashup artists can provide critical commentary on those works, expressing their perspectives on the utilized songs. As a result, mash-up can yield the first amendment expressions that the fair use doctrine was meant to protect.
>
> (Rosen, 2010)

The terms remix and sampling may also be applied to this type of "borrowing" (Chanbonpin, 2012). Here, Lankshear and Knobel (2013) note that writing satires and reintroducing old themes and content in new and clever ways is not a new practice. Indeed, they point to the fifteenth-century English playwright William Shakespeare as a master of this craft.

As these examples show, producers and users (or reusers) may disagree about what content can be owned legally, what content can be shared, and the point at which a drawing or tune can be said to belong to the public. As noted earlier in this chapter, copyright is often awarded for a finite period. It may also expire when the original work's author dies. In this instance, the work then reverts to the public domain, where the term *public domain* refers to "non-copyrighted intellectual property (such as computer software, designs, literature, lyrics, music) that can be copied and freely distributed (but not sold) by anyone without payment or permission" (businessdictionary.com, n.d.). Items in the public domain may be freely used, borrowed, and modified.

Today, borrowing and reusing is even more complicated since the internet is often described and perceived as a sharing culture. Programmers may produce and use open-source code, which may be written collaboratively by multiple producers, and produced without copyright so that others may utilize, share, and improve the code. Similarly, artists may produce visual content under a *Creative Commons license*, making it clear that their work is available for borrowing, changing, or reusing.

And it appears that not everyone feels that borrowing or sampling content is inherently ethically wrong, nor that it should be unlawful. This helps explain why enforcing copyright laws domestically and internationally has been so difficult. Noted legal scholar Lawrence Lessig believes that technological changes have necessitated new thinking about owning a piece of intellectual property. He argues that we are creating new forms of writing and literacy that may include the ability to use and manipulate various types of files – audio and video – and to put them together in new ways. He calls our attention to practices like writing fanfiction in which an author might conceive of new adventures and new types of adventures for characters who have already appeared in published literature written by another author (including retellings of stories related to *Breaking Dawn*, *Harry Potter*, or *Fifty Shades of Grey*) and photoshopping images and posting them (Lessig, 2000).

Lessig (2000) describes the old model as one in which culture is "read-only," while new culture is one in which readers can speak back and alter texts – so that it is read/written (refers to settings on old floppy discs where one could format them so that they could only be written to, versus where they could be altered). Lessig argued that traditional copyright laws choked and blocked these creative processes, since its default is "all rights reserved." He argues today for establishing a creative commons in which materials are freely available for borrowing and reuse and where the default is that you can use something rather than that you cannot.

Building an International Norm Against Piracy

As the preceding arguments show, there is not always a consensus regarding the legality or ethics of intellectual property theft in cyberspace. For this reason, some analysts are pessimistic about establishing binding international legislation that would create a uniform acceptance of the norm that intellectual property law still holds in cyberspace. In explaining why that is difficult, analysts argue that not all cultures share the same understandings of property, since these understandings come from a western legal and ethical tradition. Others argue that young people, in particular, may not share outdated or antiquated notions of ownership, having been raised with the sharing culture of the internet. Others point out that psychologically, people may not think the same way about "stealing" a computer file or image as they do about "stealing" a physical object from a friend's home. We will consider each of these arguments in detail.

Several analysts have argued that the norms and ethics governing intellectual property are not international in recent years. Instead, these ethics are based on understandings that evolved in western democratic societies over hundreds of years, including the notion of individual political rights and property rights (Ocko, 2013). Therefore, imposing these laws on nonwestern (and nondemocratic) societies through international agreements may fail because the underlying values and ethics behind these laws may be meaningless to these people in other societies.

In addition, some nations do not have a long history of respecting intellectual property and may not have the same reverence or respect for intellectual (or personal) property in their more collectivist culture. Here, Akman and Mishra (2009) argue that people's ethical decision-making is a function of their knowledge of ethics and of exposure to ethical principles, including those of their profession, the norms of their workplace, and the expectations of their workplace and their code. They argue that an employee may often be pressured by his organization to look the other way regarding intellectual property violations – such as being asked to install unlicensed software on his computer. A company may have decided to cut costs by disregarding intellectual property regimes, and the employee may be pressured to comply. Their study of IP issues in Turkey found evidence of IP violations in government and corporate organizations. The question thus remains as to whether and how successful an anti-piracy norm might be given these cultural differences (Akman & Mishra, 2009).

The International Center for Information Ethics, headquartered in Switzerland, describes three different orientations towards information production. First, they describe a European approach, which stresses the notion of authorship of a cultural product in which the author has a right to say how their information will be shared or used – and reused (for example, whether or not a piece of classical music should be used to accompany a commercial for fast food); here, the primary concern is preserving the integrity of the author's artistic vision. Next, they identify an Anglo-American approach that focuses less on artistic integrity and more on the person's economic rights whose "property" the creation is. Finally, they point to an Asian tradition that may view copying as a sign of respect for a master (International Center for Information Ethics, n.d.).

Piracy is a global problem, particularly among young people. Up to 47 percent of Americans admit to having engaged in piracy, and US college students admit to engaging in the activity routinely; most have an average of up to 800 illegally downloaded songs (Moustis, 2016). Some analysts suggest that young people simply do not subscribe to outdated privacy and private property notions, including the private ownership of ideas. Unfortunately, studies of young people, in particular, indicate that they often do not regard IP theft as an ethical issue (Hu et al, n.d., p. 127). Hu et al. indicate that people didn't believe that downloading music illegally, for example, meant that you weren't a good person. They saw the issue as unrelated to questions about their moral character. Research also shows that students do not engage in long, protracted moral searching before engaging in piracy. Instead, people tend to practice their knowledge about piracy almost immediately after learning the techniques. Over 50 percent do so less than a week after learning how to do so (Panas, 2011, p. 839),

One factor influencing students' decisions to download music may be that they don't have a direct relationship with the target of their theft, so they don't regard it as the same as stealing from a friend. They also see the likelihood of being caught as low. However, some people who engage in online piracy are part of a more significant subgroup of people who derive a psychological "high" from participating in risky behaviors, like shoplifting – both online and in person (Hardy, 2013).

And while advertising campaigns against piracy make the case that this activity is illegal and morally wrong, not everyone agrees. Indeed, some individuals and groups claim that piracy is an acceptable moral or ethical choice. Individuals making this claim may refer to piracy as a victimless crime or argue that stealing from a corporation is different from stealing from an individual since "it's only a corporation." In a blog post called "Why I Love Shoplifting from Big Corporations," Anonymous (n.d., p. 1) makes a Marxist argument, stating that he disagrees with an economic system that permits individuals or corporate entities to own goods that others may need. He feels it is fundamentally unjust for a large corporation to make millions of dollars while others are poor in that same country. He thus views his decision to enter a store and take products there to voice his objection to what he sees as a wrong economic decision. In contrast, older people are more likely to suggest that stealing is okay if a company has set its prices too high and made goods unaffordable (Babin, 1995).

Companies victimized by electronic piracy – including the Recording Industry Association of America – often mount large-scale advertising campaigns to change would-be pirates' behaviors and thought patterns. Here, they aim to change the narrative that people tell themselves and others about piracy – that it is a harmless activity. Instead, they aim to establish a norm that recognizes it as an unethical and criminal activity. These campaigns often have a simple message: Piracy is a form of theft and not a victimless crime. Indeed, some estimates are that a recording artist or studio could lose up to 20 percent of their projected profit on a project due to online piracy (Panas, 2011, p. 836).

Commercials and educational materials also seek to educate the public about the legal penalties associated with online piracy, including fines and even imprisonment. However, despite the budget for advertising of this type, the scope of the crime is growing. There is no solid or well-established *norm* against engaging in piracy, either in the United States or elsewhere.

Legal Measures Against Piracy

Furthermore, some scholars (Hardy, 2013) wonder if education campaigns are effective. They suggest that measures like increased monitoring and penalties are likely to be far more effective in preventing piracy than are measures aimed at appealing to citizens regarding the morality of their activity. While individuals may have a solid moral compass, their environments often influence them. In the case of piracy, individuals may think that the activity is wrong. However, they may still be influenced by their peers who engage in the activity and their beliefs that it is so widespread that there is little to be gained by opposing it (Hardy, 2013, p. 3). Furthermore, suppose the western intellectual tradition of private property emerged over hundreds of years. It may be unreasonable to expect similar norms to evolve over the 40 years that the internet has existed.

Nonetheless, many nations have strong laws prohibiting intellectual property theft in cyberspace. Currently, US laws are some of the strongest in the world. They include the *Copyright Act of 1976*, the *Communications Act of 1984*, and the *Piracy Deterrence and Education Act of 2003*. In addition, the FBI has run sophisticated sting operations aimed at combatting and shutting down piracy (Nasheri, 2005).

International efforts have also been mounted, mainly through the *World Intellectual Property Organization (WIPO) of the United Nations*. The World Intellectual Property Organization Copyright Treaty and WIPO Performances and Phonograms Treaties aim to combat piracy internationally and encourage nations to enforce copyright rules within their nations (Nasheri, 2005). WIPO also works to harmonize national intellectual property laws, as does the European Union Patent Office (Franklin, 2013).

In addition, the *International Federal of the Phonographic Industry (IFPI)* and the *International Anti-Counterfeiting Coalition (IACC)* work to represent the interests of companies concerned with intellectual property enforcement. The IACC puts out a list of nations warned that they are not displaying a serious enough commitment to combatting piracy. Nations on this list include Canada, India, Malaysia, Mexico, Philippines, Poland, Russia, Japan, Panama, Romania, and Turkey (Nasheri, 2005, p. 30). The IACC issues a Priority Watchlist of countries – including Ukraine, China, and Paraguay – judged as having allowed particularly egregious copyright violations.

Practitioners who work in cybersecurity are thus advised to pay close attention to the laws governing intellectual property and ensure that they observe these laws in situations where they may borrow code, content, or images. They should not assume that all content can be borrowed and pay attention to the type of license associated with particular content – which will specify whether and under what conditions content may be borrowed, modified, and shared.

CHAPTER SUMMARY

- The right to claim ownership of an idea is not new. However, new technological developments which make the reproduction of information (in written, auditory, and visual forms) easier have created new issues in IP.
- It has been challenging to establish a consensus regarding IP norms because of the architecture of the internet itself, a traditional "pro-sharing" attitude on the part of many internet users, and the fact that many different cultures with different traditions use the internet.
- Utilitarian arguments focus on IP theft's economic and intellectual costs – arguing that it makes it harder to make scientific progress if inventors can't count on being compensated for their advances.

- Virtue ethics arguments recommend that users cultivate the virtues of restraint and respect for others, even in situations where it seems easy to engage in IP theft.
- Deontological arguments ask would-be IP thieves to consider the creators' perspective of new technologies.

DISCUSSION QUESTIONS

1 **Attitudes Towards Sharing and Collaboration**
You may be familiar with many available resources for technology professionals who wish to share and collaborate in making technological progress through sites like Github and Kaggle. At the same time, throughout this chapter, we have learned about techniques like watermarking, which creators can use to ensure that they maintain ownership and control over their intellectual property.
We can think of attitudes towards intellectual property as existing on a spectrum. One end represents an open-source orientation where all materials are available to everyone without restrictions. In contrast, the other represents a total control orientation in which few materials are available as open-source, accessible products.

- Where might you place yourself on that spectrum, and why? What would be the pros and cons of each end of that spectrum?
- Is there a "happy medium' to be found somewhere in the middle of the spectrum? Can the two goals – allowing for the free exchange of ideas and collaboration, and ensuring that people get credit for their work – coexist? Why or why not?

2 **Nations and Scientists Sharing Information to Combat COVID**
Think back to the example of how nations collaborated to create COVID vaccines through "ungating" information, instead choosing to share information about the virus's genetic code freely. Think about the ethical arguments that those who chose to share made.

- Do you feel that some situations (such as the threat of a disease or national security threat) are so severe that people should be forced to share, rather than hoard, their information in those situations?
- What might those situations be, and how could you force people to do so?

3 **Using Open-Source Code**
Think back to your experiences using open-source code or visiting a site like Kaggle or GitHub when embarking upon a new project or solving a problem you encounter writing code. Think also about sites, including Reddit, where people might post a query about a coding problem they are encountering.
As you engaged in these activities, were you worried that you might be violating intellectual property rules?

4 **Working in Groups**
Think about a situation where you have shared your work or volunteered to solve a problem. Were you worried someone else might steal or take credit for your work? How do these communities police themselves and ensure that people behave ethically regarding intellectual property?

5 **IP Statements on Tech Websites**
Visit Tableau.com, Kaggle.com, and Github. Do these sites contain a statement regarding how intellectual property is defined or treated on these sites? What penalties or actions are discussed for violators of these sites?

RECOMMENDED RESOURCES

Rizzo, S., Flavio, B., & Montesi, D. (2019). Fine-grain watermarking for intellectual property protection. *EURASIP Journal on Information Security, 10*. https://doi.org/10.1186/s13635-019-0094-2.P

Tiedrich, L. J., Discher, G., Argent, F., and Rios, D. (2020). Ten best practices for AI-related intellectual property. *Intellectual Property and Technology Law Journal, 32*(7), 3–7.

US Patent and Trademark Office. (2019). Artificial intelligence: Intellectual property policy considerations. (Recordings of several talks on this subject are Retrieved June 21, 2022, from www.uspto.gov/about-us/events/artificial-intelligence-intellectual-property-policy-considerations.

Yu, D. (2018, July 30). *Intellectual property and licensing NFTs*. Half-hour video presentation on YouTube: https://youtu.be/mrsZuegV2rs

Students may also wish to visit the Recording Industry Association of America (RIAA.org) website to read their infringement, licensing, and piracy materials.

References

Akman, I., & Mishra, A. (2009). Ethical behavior issues in software use: An analysis of public and private sectors. *Computers in Human Behavior, 25*(6), 1251–1257.

Babin, B. A. (1995). A closer look at the influence of age on consumer ethics. *Advances in Consumer Research, 22*(1), 668–673.

Beckedahl, M., & Weitzmann, J. (2012). *Ten years of creative commons: An interview with co-founder Lawrence Lessig*. Retrieved June 1, 2022, from https://governancexborders.com/2012/12/18/10-years-of-creative-commons-an-interview-with-co-founder-lawrence-lessig/.

Bohannon, M. (2011). US administration's "technology netruality" announcement welcome news. *opensource.com*. Retrieved February 1, 2022, from https://opensource.com/government/11/1/us-administrations-technology-neutrality-announcement-welcome-news.

Business Software Alliance. (n.d.). *Software piracy and the law*. Retrieved June 2, 2022, from www.bsa.org/anti-piracy/tools-page/softare-priacy-and-the-law?sc_lang=en_US.

Chanbonpin, K. D. (2012). Legal writing, the remix: Plagiarism and hip hop ethics. *Mercer Law Review, 63*(2), 597–638.

Chien, S.-C. (2014). Cultural constructions of plagiarism in student writing: Teachers' perceptions and responses. *Research in the Teaching of English, 49*(2), 120–140.

Czarnitzki, D., Hussinger, K., & Schneider, C. (2015). R & D collaboration with uncertain intellectual property rights. *Review of Industrial Organization, 46*(1), 183–204.

Franklin, J. (2013). *International intellectual property law*. Washington, DC: American Society of International Law.

Hackers charged in software theft from US army, others. (2014, September 30). *Phys org*. Retrieved June 1, 2022, from https://phys.org.news/2014-09/hackers-software-theft-army.html.

Hardy, W., Krawczyk, M., & Tyrowicz, J. (2013). *Why is online piracy ethically different from theft? A vignette experiment*. Warsaw: University of Warsaw Faculty of Economic Sciences.

Himma, K. (2011). Richard Spinello and Maria Bottis: Understanding the debate on the legal protection of moral intelletual property interests: Review essay of a defense of intellectual property rights. *Ethics of Information Technology, 13*(3), 283–288.

Himma, K. (2013). The legitimacy of protecting intellectual property rights. *Journal of Information, Communication and Ethics in Society, 11*(4), 210–232.

Hockenberry, B. (2013). *The Guerilla open access manifesto: Aaron Swartz, open access and the sharing imperative*. Rochester, NY: St. John Fisher College Fisher Digital Publications.

Hu, Q. Z., Hu, Q,, Zhang, C., & Xu, Z. (n.d.). Moral beliefs, self-control and sports: Effective antidotes to the youth computer hacking epidemic. In *45th Hawaii international conference on system sciences*. Retrieved June 1, 2022, from http://ieeexplore.ieee.org/document/6149196/.

International Center for Information Ethics. (n.d.). *The field*. Munich: International Review of Information Ethics.

Lankshear, C., & Knobel, M. (2013). Digital remix: The art and craft of endless hybridization: Keynote. In *International Reading Association Pre-Conference Institute "Using Technology to Develop and Extend the Boundaries of Literacy"*. Toronto: International Reading Association.

Lessig, L. (2000). *Code: And other laws of cyberspace*. New York: Basic Books.

Lindsay, J. (2014, February 11). Hacked medtronic, Boston Scientific, St. Jude networks suffer cybersecurity breaches. *Med Device Online*. Retrieved June 2, 2022, from www.meddeviceonline.com/doc/hacked-medtronic-boston-scientific-st-jude-networks-suffer-cybersecurity-breaches-0001.

Maisel, Y. (2014, May 16). A call for open access to medical journals for rare disease "detective" patients and advocates. *Global Genes.org*. Retrieved June 2, 2022, from https://globalgenes.org/2014/05/16/a-call-for-open-access-to-medical-journals-for-rare-disease-detective-patients-and-advocates/.

Mancic, Z. (2010). Cyberpiracy and morality: Some utilitarian and deontological challenges. *Filosofia i Drusto*, 103–117.

Moore, A. D. (2005). Introduction to information ethics: Privacy, property and power. In A. D. Moore (Ed.), *Information ethics: Privacy, property and power* (Retrieved from Unsworth, Kristene and Moore, Adam D., "Introduction" to Information Ethics: Privacy, Property, and Power (September 28, 2005). Information Ethics: Privacy, Property, and Power, p. 11, A. Moore, ed., University of Washington Press, 2005).

Moustis, J. Jr., & Root, A. (2016). Curing Americans Kleptomania: An empirical study on the effects of music streaming services on piracy in undergraduate students. *Timelytech.com*. Retrieved January 2, 2021, from http://illinoisjltp.com/timelytech/curing-americans-kleptomania-an-empirical-study-on-te-effects-of-music-streaming-services-on-piracy-in-undergraduate-students-2/.

Nasheri, H. (2005). *Addressing the global scope of intellectual property law (final report)*. Washington, DC: NIJ, The International Center (US Department of Justice).

Nezami, K. (2017, July 19). AI and intellectual property: Can AI be an inventor? *IP Harbour*. Retrieved from https://ipharbour.com/blog/ip/can-ai-inventor-part-1/.

Ocko, J. (2013). Copying culture and control: Chinese intellectual property law in historic context. *Yale Journal of Law and Humanities, 8*(1). Retrieved June 1, 2022, from http://digitalcommons.law.yale.edu/yjlh/vol8/iss2/10.

Panas, E. A. (2011). Ethical decision-making in electronic piracy: An explanatory model based on the diffusion of innovation theory and theory of planned behavior. *International Journal of Cyber Criminology, 5*(2), 836–859.

Pinkos, S. (2005). *United States patent and trademark office, piracy of intellectual property: Statement of Stephen Pinkos, deputy undersecretary of commerce for intellectual property*. Washington, DC: United States Patent and Trade Office.

Rosen, B. E. (2010, February 4). Mashup: A fair use defense – by "Ryan B". *Yale Tech Law Blog*. Retrieved June 4, 2022, from https://yalelawtech.org/2010/02/04/mashup-a-fair-use-defense/#:~:text=For%20this%20hypothetical%20fair%20use%20defense%2C%20let%E2%80%99s%20delve,however%2C%20samples%20are%20usually%20meant%20to%20be%20recognizable.

Schultz, A. (2016, January 25). SAT integrity falls victim to China cheating scandal. *Barron's*. Retrieved June 1, 2022, from www.barrons.com/articles/sat-integrity-falls-vivtim-to-china-cheating-scandal-1453713163.

Shackelford, S. (2016). Protecting intellectual property and privacy in the digital age: The use of national cybersecurity strategies to mitigate cyber risk. *Chapman Law Review, 19*(2), Retrieved June 5, 2022, from https://papers.ssrn.com/sol3/papers.cfm?abstract_id=2635035.

Spinello, R. A. (2003). *Cyberethics: Morality and law in cyberspace* (2nd ed.). London: Jones and Bartlett Publishers International.

Tavani, H. (2005). Locke, intellectual property rights and the information commons. *Ethics and Information Technology*, 7(1), 87–97.

Urban, G. (2015). Complicity in cyberspace: Applying doctrines of accessorial liability to online groups. In R. G.-L. Smith (Ed.), *Cybercrime risks and responses: Eastern and western perspectives* (pp. 194–206). New York: Palgrave Macmillan.

Varelius, J. (2014). Do patents and copyrights give their holders excessive control over the material property of others. *Ethics and Information Technology*, *16*(1), 299–305.

Walker, D. (2015, April 2). Man pleads guilty to intellectual property theft conspiracy impacting microsoft, other firms. *SC Magazine*. Retrieved June 1, 2022, from www.scmagazine.com/hacking-ring-member-pleads-guilty-to-st.

Part III

7 Ethics of Artificial Intelligence

LEARNING OBJECTIVES

By the end of this chapter, students will be able to:

1. Define critical terms including artificial intelligence, machine learning, general and superior artificial intelligence, and singularity
2. Describe ethical issues associated with the granting of agency to artificial intelligence
3. Describe methods for establishing human responsibility for AI decision-making processes, including the concept of meaningful human control
4. Describe ethical arguments related to the value of authenticity as an obstacle to further development of artificial intelligence

- In spring 2018, a news story emerged regarding an experiment at Elon Musk's OpenAI Lab. In the experiment, AI bots solved a problem that involved playing a game in which they bid on objects. At some point, the bots began behaving in unexpected ways. Specifically, they created a unique language and began communicating in a language unintelligible to human researchers. After a few exchanges, researchers shut down the experiment, nervously joking about the singularity (Tangermann, 2017).
- In spring 2021, an Israeli military operation called Guardian of the Walls took place. During this event, the Israeli intelligence corps' elite Unit 8200 created an algorithm and code for three new programs – Alchemist, Gospel, and Depth of Wisdom. The Gospel program used AI to generate recommendations for troops and military intelligence, including targeting information. The Israel Defense Force's advanced AI Technological Platform was also used to aggregate data on terrorist groups in the Gaza Strip. Big Data was also used to map Hamas's underground network of tunnels to get a picture of the depth and thickness of the tunnels and the nature of the routes used (Ahronheim, 2021).
- The Chinese company Yitu, founded in 2012, created the Dragonfly Eye image platform containing more than 1.8 billion photographs. Yitu claims that the platform's facial recognition and image recognition algorithms can identify any individual in a database within three seconds. The platform includes images from the People's Republic of China national database, plus photos taken at borders. The platform was valued at $2.4 billion in 2018. The facial recognition algorithms developed by Dragonfly Eye have won awards from the US National Institute of Standards and

DOI:10.4324/9781003248828-10

Technology and the US Intelligence Advanced Research Projects Activity (IARPA) (Feldstein, 2019).

- Increasingly, we encounter and interact with many online entities through our automated programs. As Gent (2020) notes:

> if you have been online, it is almost impossible for your experience not to be shaped by bots. These automated chunks of code preprogrammed to perform a certain task repeatedly account for as much as 39 percent of activity on the web.

The "army of bots" online today includes web crawlers, or "spiders," which perform searches and collect information; monitoring bots that periodically check websites for bugs, poor performance, and outages; moderator bots that automatically flag inappropriate content and hide it; and spiders that look across the web for copyright infringement and stolen content. Trading bots interact with financial services, including the stock market, while scraping bots collect information, sometimes sharing the information they collect with spambots. A recent report suggests that most false health information is generated by automated accounts (bots) and individuals (trolls) "who misrepresent their identities to promote discord" (Karami et al., 2021).

- In 2017, Russia's president, Vladimir Putin, famously said that whatever country becomes the leader in artificial intelligence "will become the ruler of the world" (Petrella et al., 2021).

But what do these five stories – about robots talking to one another in a language unintelligible to humans, about the use of algorithms to make strategic decisions during wartime, about algorithms used to identify people's faces in an online database, about armies of bots impersonating humans online and about a possible AI "arms race" between Great Powers have to do with cybersecurity ethics?

First, these stories illustrate the ubiquitousness of artificial intelligence. Today, artificially intelligent bots and algorithms play a part in many different social interactions between people – making economic, political, and military decisions and acting in ways that have real-world implications for the lives of real people. Our online ecosystem is increasingly defined and shaped not only by human actors' actions but also by pieces of code and algorithms, which often act autonomously or without human monitoring. And these stories show that we as humans may not always be aware that we are interacting with artificially intelligent entities acting autonomously.

These stories also illustrate how artificially intelligent agents are often better and faster at carrying out many jobs – from reading medical scans, to identifying people in photos, to solving complex math problems – than humans are. Humans are increasingly reliant on AI to perform tasks too complex or large (or expensive) for humans to do. We can say that many jobs formerly done by humans have been outsourced to AI.

But these stories also highlight some of the dangers and risks of relying on artificial intelligence. The Russian president was not just concerned with having Russia develop artificial intelligence. He also wanted to make sure that Russia was developing better artificial intelligence – more reliable, more secure, and less vulnerable to cyberintrusions – than the technology that his peer competitors (like the United States and China) were developing. The fact that so many sectors of a developed economy now depend on artificially intelligent agents to carry out some aspects of their jobs means that new types of security vulnerabilities are emerging – which presents new challenges for those in the field of cybersecurity. An attack on the algorithms that engage in trading currency and cryptocurrencies on Wall Street and the world markets can potentially crash economic systems domestically and internationally.

And some legal analysts are concerned that establishing responsibility could be particularly challenging when a programmer's mistake or error in the ML model creates such a scenario. An attack on artificial intelligence systems has been described as a "perfect storm" of events, since it might lead to multiple different outcomes. It could also create a situation where establishing blame or responsibility could be difficult since we don't always perfectly understand what this algorithm (to whom we have outsourced so much power) is doing. How do they work, and how do they make decisions? In many instances, their exact mechanisms seem opaque and mysterious.

Cybersecurity Issues Related to Artificial Intelligence

Thus, we can see how the brave new world of AI raises new questions (including ethical queries) for those engaged in cybersecurity. In this chapter, we will first consider the question of agency. Agency refers to the ability of an entity to act or take action. Philosophers generally agree that to act agentically, one must act intentionally and with reason. That is, falling down the stairs accidentally is not an example of acting agentically, while stepping off a diving board would be (Schlosser, 2019).

Someone acting agentically (or with intention) can describe why they did something and give an account of their actions. A human reading an x-ray might say they want to help a patient. A soldier taking action to kill another human in combat might say that he wishes to save his country from invasion. In addition, some philosophers distinguish between the agency displayed by humans and how machines operate by referring to the **Theory of Mind (TOM).** Humans, they argue, have a highly developed Theory of Mind. We understand when interacting with others and can consider how others perceive our actions and are likely to react. For example, as a human, your reason for acting might be (and often is) the desire to impress or please another human. Robots, in contrast, lack a Theory of Mind. As Shorey (2016) argues, the robotic home helper Alexa may respond to your request to play a song by Beyoncé, but Alexa is not doing so to please you. Alexa will respond equally well to any human and carry out any command unless her program explicitly forbids her.

Thus, a philosophical and ethical question arises when actions typically undertaken by humans are instead "outsourced" to nonhuman entities like a computer program or an algorithm. An artificially intelligent computer program can "learn" to act or carry out actions or activities by being trained and fed information. After a bot, for example, assimilates information from a database of images, it will be able to identify a photo. However, it could neither explain why it was doing what it was doing nor give an account of itself and its actions. It would not understand the moral import of its actions if its identification or misidentification of an individual resulted in their going to prison. It would not feel guilt or remorse.

Thus, the question is, "If bots cannot and do not act agentically, then who or what is ethically and morally responsible for their behavior? What are the moral and ethical responsibilities of creating these programs?" Many traditional cybersecurity tasks have now been outsourced to AI – from identifying security breaches to responding to systems attacks, including hacking back. But where does the line lie between what only humans should and can do, what only AI can and should do, and situations where both must cooperate and work together?

Are some tasks too ethically or morally sensitive to be simply handed to a bot or algorithm to carry out? If so, which tasks? Who should be held responsible when a task has been outsourced and an unfavorable outcome ensues? Is it enough to build a secure system and then let it operate on its own, or does the engineer also need to think about how to continually monitor a system as it works and possibly even evolves, particularly with the growth of AI? This chapter introduces the notions of "humans in the loop," meaningful human control (MHC), and how a process can be interrupted or taken off-line if it behaves unexpectedly. Today, a bot can diagnose and repair itself – but should it?

This chapter also introduces ethical and moral thinking about trust, including what it means to trust another person or technology and the inherent risks. Finally, we will consider emerging legislation, including the European Union's Strategy on Artificial Intelligence, focusing on the ethical values presented in this document.

Defining Artificial Intelligence

What exactly is artificial intelligence? The website for International Business Machines, or IBM, describes artificial intelligence as a field combining computer science and robust datasets to enable problem-solving. Most analysts agree that artificial intelligence refers to the application of technologies that allow computers to carry out tasks that usually require human cognition (Sanchez-Herrero, n.d.) – such as plotting a route from one location to another, searching for the best price on a good to be purchased, or scanning resumes for keywords to find the best candidate for a job.

Such tasks can range from very simple to more complex tasks. We can think of such tasks as existing on a spectrum, from those requiring the one-step application of a simple rule (like counting how many letters are in a word or alphabetizing a list of names) to those requiring more complex multi-step problem-solving processes that mimic human cognition.

In thinking about artificial intelligence, practitioners commonly refer to the work of Alan Turing, a British computer scientist. He contributed to problem-solving and code-breaking for British intelligence during World War II. Turing described the possible future advent of a machine so adept at mimicking human thought processes and solving problems that people would be unable to distinguish this "intelligence" from an actual human. (Subsequently, this standard has become known as the "Turing Test.") Thus, Turing's vision rested on a computer system capable of mimicking what a human does – thinking, making observations, and solving problems like a human. This capacity has come to be known as *general artificial intelligence.*

To "teach" a robot to carry out more sophisticated tasks, engineers utilize *machine learning* programs in which machines are "trained" to perform a task through repetition. Machines are fed datasets that show how a particular rule is applied or how a particular task is performed (i.e., a machine might learn how to triage patients in an emergency room by assimilating all of a hospital's patient data gathered over several years). As a result, they "learn" – deriving algorithms to understand how rules are applied and eventually learning to apply the rules themselves. In many instances, machines can go beyond human cognition and human practices to perform new tasks faster and more efficiently than humans.

We use the term *superintelligent AI* to describe situations in which artificial intelligence has gone beyond human cognition; for example, utilizing more variables in analysis than a human could track or work faster with larger datasets. Superintelligent artificial intelligence is often generated through "deep learning" in which computers are organized into a neural network that resembles a human brain. Information is then circulated along pathways between nodes, allowing a program to connect information and decide the best way to process information between nodes. You may be familiar with Google's DeepMind, a machine learning program based on a neural network. Such machines may create their unique methods for solving problems and may work faster and more efficiently than humans do, but they often act in incomprehensible ways. In situations like this, we can begin to speak of ceding authority or agency to an artificially intelligent agent or even depending on or trusting such an agent.

Some futurists warn of a day in which such computers could suddenly become much more intelligent than humans and warn of a future in which humans might find themselves living in a world where the rules are written not by humans but by AI, and where the world is perhaps organized in a way which does not make sense to us humans. The term *hard takeoff* or *intelligence explosion* refers to a situation in which computers might suddenly acquire this new capacity. Some writers also refer to *the singularity*, meaning a period in which computers might become able to communicate with one another or even achieve sentience, or the capacity to perceive themselves, feel, and act with agency.

As we have seen, then, artificially intelligent programs have the potential to revolutionize how we practice our occupations and make new knowledge in many fields, from education to health care, to city planning to finance, to the legal sphere to issues of justice and equity. But it also raises an ethical issue.

In this way, we humans have a curious relationship with AI. We are in many ways dependent upon it, but we do not trust it either. We find ourselves increasingly likely to offload duties and responsibilities to AI. Still, at the same time, we worry that AI may make mistakes, including those we do not catch or understand.

It can be said that we do not trust AI. For this reason, we humans have set about developing safeguards for the activities of AI. While we have noted that AI systems are incapable of acting agentically (with reason and intention), they may still act autonomously. For example, a weapon might be created to scan an environment, identify suspected terrorists using photo matching and facial recognition technology, and then fire a weapon at the suspected terrorist. Such a weapon would be described as *fully autonomous*, or in which the human was "*out of the loop*."

Autonomy and Humans in the Loop

However, autonomy resides on a spectrum, with products operating from semi- to fully autonomous. In some instances, we have created systems that allow humans and machines to work together in concert, acting as a hybrid human and machine interface. For example, today, a military fighter might use a weapon that relies on artificial intelligence to view the battlefield and precisely target a rifle shot more accurately and quickly than a human. However, the AI is not permitted to fire the weapon. Only the human warfighter can undertake this action. Such as system is referred to as one in which there is a "human in the loop."

The decision to create artificial intelligence products with varying levels of autonomy or human guidance rests on the understanding that some decisions are perhaps too important to be left only to an algorithm due to the significance of the outcome of that decision. The European Union coined the term *high-risk artificial intelligence* to refer to decisions that might have life-altering repercussions. European Union regulations require that such high-risk AI decisions (such as the decision to deny someone a loan, a job, or an insurance policy) must ultimately be transparent (or capable of being understood by those who are subject to such decisions) and that humans would tightly monitor such decisions to make sure that they were occurring in a fair and unbiased manner (Sparkes, 2021).

Box 7.1 Going Deeper: The European Union Artificial Intelligence Act

In April 2021, the European Commission proposed the Regulation on Artificial Intelligence, the EU AI act. As of this writing, this act was in the discussion and negotiation phase, with amendments being proposed to the initial draft; however, it was likely to be passed in 2023.

Like most European Union legislation, this act aims to harmonize regulations and policies throughout the European Union. Due to its comprehensive scope, it may, as the Budapest Convention on Cybercrime and the EU General Data Protection Regulation (GDPR), serve as a template for other non-EU nations in the future. As Edwards notes, this act is not aimed primarily at regulating the use of artificial intelligence by platforms like Facebook, since other legislation within the European Union already does that. Instead, this legislation aims to regulate the use of artificial intelligence by the public sector and law enforcement agencies.

While the GDPR regulates how your data is collected and stored by actors, including the government, the EU AI act aims to regulate how this data is used and stored (and by whom) for forecasting, using algorithms to make predictions about your credit risk and the likelihood that you would engage in criminal behavior and so forth. In this way, the new legislation overlaps with the GDPR. And as is the case with the GDPR, in applying the EU AI act, other non-EU corporations wishing to do business in the EU will also be subject to these regulations.

One unusual feature of the EU AI act is that the act is written to regulate the behavior of providers of AI-enabled systems and, in some cases, the users of these systems. System providers incur new obligations, including the obligation to conduct the necessary security audits and assure that systems and products comply with EU regulations. Thus, for example, if a European company purchases a subscription to software as a service (SaaS) from a provider like Blackboard or Salesforce, and the product contains analytics or machine learning components, these components must comply with EU directives. Sometimes, a company purchases software and then configures or modifies it in-house or works with a vendor for specific needs. In this case, the company may also be subject to specific new regulations. A user might also be a municipal school system or police force that reconfigures or modifies a system.

Some analysts are concerned that the act is too broad – assigning too many regulatory responsibilities to too many different types of actors since both the terms provider and user of AI-enabled software are new terms and may not be legally defined.

Another unusual feature of the act is that it breaks AI-enabled activities into risk groups. Some AI-enabled activities (like configuring and using a spam filter) are considered relatively low risk. In contrast, others, like using AI programs to make decisions about employment, education, justice, and immigration are considered higher risk. The act also emphasizes transparency – insisting that AI-enabled chatbots (such as those found on a shopping or customer service website) should be labeled so that consumers are aware that they are not interacting with a person.

Finally, the act explicitly forbids companies from engaging in certain AI-enabled activities, deeming them as posing an unacceptable societal risk. These activities include the creation and application of social scoring algorithms, the use of facial recognition software for specific applications, and the use of techniques such as subliminal messaging (Edwards, 2022).

Bibliography

Edwards, L. (2022). *The European Union artificial intelligence act: A summary of its scope and significance*. London: The Ada Lovelace Institute.

The Ethics of Trust

What does it mean to trust a machine, and how is this different from the trust you might place in another human being? What does it mean to "trust" someone or something? Many philosophers describe trust in utilitarian terms; trust is a rational action or decision people make to maximize efficiency and utility. Trust is about delegating action to someone or something else. You then take a risk or a gamble by "trusting" that person to act on your behalf, often saving you time and trouble.

For example, we may trust our local meteorologist to tell us the weather rather than undertaking our research. We save time and energy but have to reckon that the weather report might be wrong, and we may find ourselves at an outdoor event without an umbrella. Note that the decision to "trust" the weatherman has nothing to do with whether or not we like him. We are undertaking a rational action to save time, not forming a friendship with the individual (Hardin, 2006).

Often the consequences of trusting someone in this way are pretty small. We won't die if we get wet in a rain shower. But we might overtrust a shady used car salesperson and get talked into buying a poorly performing used car. Some trust decisions are risky and may lead to a negative outcome.

Ethicists thus talk about ways in which we can build trust. We often build more trust towards an institution or individual throughout a long relationship (such as a bank we have banked with for several years). We might trust individuals or groups who look like us or share our values more. We might have more trust in an established technology than one which exists in a beta version (Taddeo & Floridi, 2017)

But how do we apply these understandings about trust to relationships between humans and machines, including autonomous agents? When are autonomous agents presumed trustworthy agents, and can a human establish a trusting relationship with an autonomous agent? Specifically, when should a warfighter trust the information provided by a machine or via electronic communication, and what are the risks and dangers posed due to threats to the trust relationship?

Cybersecurity practitioners know the risks of trusting a computer program or an AI algorithm. The program might contain malware or have been fed incorrect or improper data. In August 2017, a GPS failed to guide an American naval ship, leading the ship to become lost in Russian waters (in this case, Russian operatives had hacked the GPS) (Hambling, 2017).

Cybersecurity practitioners thus need to consider the attribution problem (or ensuring that the entity with whom you interact is appropriately identified and is who you believe they are). Authorization techniques can be used to tackle this problem.

They also need to consider problems related to communication and competence (that is, making sure that the AI understands the commands it is given and has the resources to carry out the command). The reason why policymakers strive to make the decisions made by AI transparent to those affected by this hem is that people are more likely to trust AI if they understand how it works, including what biases might exist and how they have been identified and corrected for. As noted earlier in this text, the subjects of AI enabled decision-making to want to be assured that they are being treated fairly and equitably (Henderson, 2019).

We can also consider the problem of motive. In "outsourcing" the weather forecasting to our local meteorologist, we share the same motives as the meteorologist. The weather person and her listeners want to be dressed appropriately for the weather and prepared for what might occur. But can an AI system we trust to drive our car, hire an office assistant, or decide about a bank loan be said to share the same motives we do in assigning the task? Here we again encounter the problem of agency. The AI doesn't "want" anything – except perhaps to solve a problem in the most efficient way possible (quickly, using the most efficient ways of working with data and the least amount of energy and electricity, or storage space, for example). Therefore, asking an AI-enabled algorithm to decide how long a prison sentence an individual should receive may be inappropriate. An individual making a sentencing decision would be making this decision within the context of such values as equity and justice, which would be irrelevant to an algorithm. And as noted in the previous chapter, even teaching an AI to avoid making biased decisions is not the same as teaching it to be merciful or a lover of justice. You may feel that this distinction matters.

It is important to consider ethical issues related to motives due to our growing willingness to cede decision-making authority to artificially intelligent agents – a phenomenon sometimes referred to as "technology paternalism." Today, we might let a diet or nutrition program plan our meals, for example, or train for a marathon by using a fitness program that plans our workouts. The assumption is that the platform or utility "knows what is best for us." But Royakkers et al. (2018) ask about a future in which, perhaps, we might see our artificially intelligent agents acting autonomously based on what they assume they know about our wants and needs. We can envision an artificially intelligent refrigerator that puts us on a diet by limiting our access to specific sections of the refrigerator or even a dating application that steers us towards certain types of partners and away from others. Royakkers et al. (2018) warn that "the implicit enforcing or provoking of certain behavior can endanger personal autonomy." They suggest that citizens should be informed when they encounter persuasive technology. However, a utilitarian might argue that if the artificially intelligent agent prevents people from drinking and driving by controlling people's vehicle access when they are inebriated, the utility generated is more significant than any possible threats to our autonomy.

Earlier, we introduced the notion of a "human in the loop" to describe situations in which human intervention is needed as an augmentation or control over the actions of an artificially intelligent agent. Royakkers et al. (2018) argue that man is "out of the loop" in many situations today, as in situations where persuasive technology guides or steers our actions without our awareness or consent.

In addition to the problem of motive, some practitioners believe that we can never fully trust an artificially intelligent agent because of the problem of commitment. Because the artificially intelligent entity is not human, it will not suffer the same consequences from being wrong in its assessment that a human might. Consider a situation in which your GPS has given you wrong driving directions. Perhaps your GPS is not programmed to know and remember which roads flood when there is a lot of rain. Therefore, you might know as a user not to "trust" your GPS during the rainy season. Ultimately, it doesn't matter to your GPS whether or not you receive the wrong driving directions. If the directions are inaccurate, you will suffer the consequences (the destruction of your car's engine or even drowning), but the GPS itself will not. For this reason, some institutions, like the US military, are building redundancies into systems like a GPS-enabled navigation system. Naval recruits still learn how to calculate geographic positions by hand. If such technology is unavailable or inaccurate, commercial airline pilots still learn how to land a plane without electronic navigation. Ultimately, we are uncomfortable ceding this much authority – or imbuing that much trust – in a technology.

Box 7.2 Tech Talk: Cybersecurity, Ethics, and Self-Driving Cars

Does the car you drive have features like cruise control and lane assist? If so, then your car represents the first wave on the path towards the eventual widespread use of fully autonomous vehicles. An autonomous vehicle operates and performs tasks independently of the driver or under its power.

The Society of Automotive Engineers currently recognizes five levels of autonomy or automation within the autonomous vehicle industry. Level 1 represents the lowest level of autonomy, "assisted automation," and refers to cars that might include features like lane assist but where a human driver is still entirely in control of the vehicle. Level 3 represents a system where the human and the system act as a sort of hybrid, with the human checking in on the driving system but allocating many decisions to the system itself. Finally, Level 5 represents full autonomy, where the vehicular system makes decisions on its own (Taeihagh & Lim, 2019). Such a vehicle might, for example, choose a route based on its superior driving skills and excellent command of geography – in comparison to a human driver – honed through artificial intelligence processes such as machine learning. A level 5 system might also alter the route in response to new information about obstacles, such as traffic jams.

Fully autonomous or self-driving vehicles rely on systems that can integrate and communicate information. Tools for gathering information can include cameras, sensors, GPS, radar, and onboard computers. Systems must share data within the vehicle itself to drive your car forward, and data must also be shared in a vehicle-to-vehicle relationship. Finally, data must be shared between the vehicle and its surrounding environment. In the future, we may also develop both intelligent roads and intelligent roadside units that will collect data about traffic patterns and the vehicles using the roads and then share them with vehicles as they pass. In addition, autonomous vehicle guidance systems may interface with other information sources along their route – integrating traffic data from a service like Waze, receiving messages about road closures and warnings about road accidents from a service like a state or local road system exchanging information with other vehicles. For example, your vehicle might pick up information from other vehicles about their route and destination and receive warnings before passing another vehicle. Thus, driving a self-driving vehicle requires creating, maintaining, and securing an entire data ecosystem (Rasheed et al., 2020).

The Benefits of Self-Driving Vehicles

The upside to a transition to autonomous vehicles is clear. Perhaps most importantly, automated programs can save lives by reducing the number of road accidents caused by human error factors (such as a sleepy, alcohol-impaired, or elderly driver misjudging a distance on the road). Currently, it is estimated that 90 percent of auto accidents are the result of human driver error. Such systems can also increase mobility and independence for individuals with disabilities and the elderly who may no longer feel confident in driving themselves.

And fully automated vehicles often perform in more fuel-efficient ways, saving more miles per gallon and therefore saving drivers money and helping to reduce our reliance on fossil fuels and saving the environment. By exchanging information, self-driving vehicles may also reduce traffic bottlenecks and congestion, helping people arrive at their destinations faster and more efficiently.

The Downside

But for all of the upsides cited here, there are also risks. As Raijin points out, all of the data streams created and exchanged on our highways and cities create new security and privacy vulnerabilities. He writes that a "lack of information security can lead to criminal and terrorist acts that eventually cost lives" (Raiyn, 2018).

Specific cybersecurity threats include the possibility that malicious actors could hack into data streams, corrupting data and potentially causing fatal accidents or socially disruptive large-scale pileups. An attacker might even take control of a vehicle and cause it to crash into an obstacle or another vehicle. Thus, the information provided within the vehicle, in information loops between vehicles, and between the environment and the vehicle must be consistent and reliable. In addition, vehicles and their systems must be able to quickly detect intrusions into these loops and perform corrective actions through the development of solid intrusion detection algorithms.

In addition, self-driving automobiles rely on GPS data. Hackers might also attack space satellites that provide this data – potentially crippling the transportation and related industries, like shipping, manufacturing, and agriculture. Attackers might also use GPS spoofing, causing drivers to get lost or even experience harm by entering a flood zone or driving the wrong way on the highway.

Finally, some analysts worry about how passengers and drivers might lose their privacy. In operating such systems, data on people's driving activities is compiled, shared, and often linked to specific individuals and vehicles. Such information could potentially be used in criminal proceedings, to embarrass individuals, or to engage in blackmail or extortion.

Currently, most developed systems operate according to the model by which vehicle owners do not own the navigation systems that would drive their vehicles. Instead, vehicles rely on software that vehicle owners subscribe to (using a Software as a Service model). When users download the software and register their accounts following the purchase or rental of their vehicles, they sign a Terms of Service agreement which gives the software provider many rights regarding their data – including the ability to save and store this data, as well as the ability to monetize this data through, for example, sending ads to drivers about places they may be traveling to. Many potential customers may resist switching over to self-driving vehicles unless strict privacy laws and regimes are developed, governing who may access information about their driving activities, for what purposes, and under what conditions. In addition, regulations need to be developed governing how long such data is stored and how it will be destroyed.

Other potential ethical issues include issues related to responsibility and liability in, for example, the event of a motor vehicle accident. Could a self-driving car (or its software) be held responsible for causing an accident, and could the software company be liable for either criminal penalties or civil damages?

Here, a related debate among ethicists asks whether autonomous systems will ever be capable of "thinking ethically" if, for example, a pedestrian steps in front of a self-driving car. Should the system's primary goal here be to protect the safety of its passengers, even if doing so necessitates running over the pedestrian, or should the vehicle seek to save the pedestrian's life even if doing so might injure the passengers within their car? Because artificial intelligence is not aware or capable of examining its actions and their import, many ethicists believe an autonomous vehicle system could only implement a very rudimentary sort of "ethical calculus," based on utilitarian ethics in which the system would weigh the costs and benefits of adopting each course of action and then choose the one which creates the least amount of harm.

Combatting Cyberhacking

Security engineers have identified several different paths for ensuring data security in the autonomous vehicle arena. Developing secure systems for authentication is a fundamental principle here. As vehicles exchange data streams, a system must allow each vehicle's data to be linked to their identities, ensuring that outsiders have not entered the system to provide corrupted data. Raijin suggests that biometric data like an iris scan might be used, allowing a driver to link her profile to a vehicle and exchange this information with other vehicles. In addition, many engineers hope to build redundancies into systems such that a vehicle, for example, might receive multiple inputs about its location or the activities and locations of other cars on the road, such that information received from one source can be checked against information received from other sources, assigning a reliability rating to the information.

Competing Values

As Taeihagh and Lim (2019) suggest in their analysis of international legal regimes governing self-driving vehicles, different nations are at different points in moving towards adopting these vehicles. They may also have different views about what matters in regulating these vehicles. They note that, not surprisingly, the European Union's proposed legislation puts a significant emphasis on the rights of drivers and passengers to have privacy about their data. In contrast, proposed legislation in American states often emphasizes a limited government role which aims to increase competition in this rapidly growing field.

Bibliography

Raiyn, J. (2018). Data and cybersecurity in autonomous vehicle networks. *Transport and Telecommunication*, *19*(4), 325–334.

Rasheed, I., Hu, F., & Zhang, L. (2020). Deep reinforcement learning approach for autonomous vehicle systems for maintaining security and safety using LSTM-GAN. *Vehicular Communications*, *26*(1). Retrieved June 3, 2022, from https://doi.org/10.1016/j.vehcom.2020.100266.

Taeihagh, A., & Lim, H. (2019). Governing autonomous vehicles: Emerging responses for safety, liability, privacy, cybersecurity. *Transport Reviews*, *39*(1), 103–128.

Creating Trustworthy AI

Many cyber-attack strategies that concern cybersecurity experts in safeguarding AI are explicitly aimed at degrading our trust in AI-enabled processes. How might an adversary interfere, for example, in a machine learning process to trick the AI into learning the wrong lessons? Ma et al. (2019) describe how "adversarial machine learning occurs." Data poisoning is the term given to an attack where the attacker attempts to manipulate the machine learning model. The AI learns the wrong lesson (such as learning to label spam e-mails as safe). More general model inversion attacks occur when an adversary accesses a machine learning model to access the materials used to train the artificial intelligence. As a result, sensitive information about individuals might be revealed.

Adversaries may also attempt to "map" existing AI structures such as neural networks used for machine learning. The aim here is often to engage in military or commercial espionage through reverse engineering. If an adversary can understand how a system is configured, this information can be used to build one's model (Free & Zhang, n.d.). (Such techniques are not unique to the field of artificial intelligence. America's adversaries have used reverse engineering in the past to steal technology used to build battleships and nuclear weapons.) This way, any "*first-mover advantage*" the original builder might have is quickly undermined. As noted earlier in this text, such actions would violate existing intellectual property rights law nationally and internationally and, in some instances, might even constitute an act of war. Nonetheless, in the high-stakes "Artificial Intelligence Arms Race," such activities are likely ongoing.

Artificial Intelligence, Artificial Environments, and Ethics of Authenticity

Another ethical issue concerning artificial intelligence is the issue of authenticity, particularly when artificial intelligence is used to generate materials like "deep fakes" or videos that are altered to make it appear that events have occurred when in fact, they have not. The *Oxford English Dictionary* defines "authentic" as original or faithful to the original. As Guignon (2008) writes: "To say that something is authentic . . . Is to say that it is what it professes to be, or what it is reputed to be, in origin or authorship."

This issue became salient to those in the tech community in July 2021, when a documentary was released about the life of the famous chef and travel writer Anthony Bourdain. The documentary's readers heard a computer-generated voice matching the late writer's. It appeared that Bourdain himself was saying and reading things he never said.

Applying the three paradigms of utilitarianism, virtue ethics, and deontological ethics here, we can ask whether creating an "artificial or inauthentic Anthony Bourdain" to narrate the documentary represents an ethical breach.

A utilitarian might argue that no one has been harmed by exposure to this inauthentic content. Indeed, one could argue that including this material has enhanced viewers' experiences. For example, at one point in the documentary, Bourdain "reads" from his journal, allowing the reader to glimpse his psyche. As Bourdain struggled with depression and committed suicide in 2018, the documentary viewer might thus better understand Bourdain's struggle by "listening" to him read from his journal. Here a utilitarian might argue that this positive act of dubbing a voice into a documentary is thus fundamentally different from, for example, selling someone counterfeit goods.

Furthermore, one might argue that using a "deep fake" video could be equally helpful. For example, if the population was panicking after a terrorist event in Washington, D.C., a "fake" video of the president addressing the population, instructing them that everyone was safe and that they should be calm, could avert chaos and societal breakdown. Wouldn't this positive good produced far outweigh any qualms one might have about using an inauthentic video? A utilitarian might explain that these actions are fundamentally different from unethical acts like theft or lying.

But from a deontological ethical standpoint, we can ask if presenting inauthentic material represents a type of manipulation. The material producer is aware that they are deceitful in failing to mention that the presented material is inauthentic. The deontological lens would require a producer of such material to ask themselves, "How might I feel if I were not the fake producer but rather the viewer? Would I wish to be deceived in this way?" Here we can consider how we feel when we call a phone number like customer service and are perhaps taken in by a computer-generated voice that we assume is human. Here ethicists have asked if it would ever be appropriate to, for example, hire a robot caregiver for an elderly loved one, particularly if the person being cared for believed that their caregiver was a human. Is there something inherently manipulative or deceitful in such a scenario? (Blackford, 2012). Is allowing people to develop feelings for or relationships with artificially intelligent agents they believe to be human inappropriate and perhaps even wrong?

Finally, virtue ethics describes authenticity as a virtue, with the opposing vice being one of inauthenticity or something false or fake. Someone who wished to embody the virtue of authenticity might then say that no circumstances justified passing off inauthentic content as accurate – not even the national security scenario described earlier. However, as noted in our chapter on privacy, a counterargument would state that no one is completely transparent all the time, nor is anyone entirely honest. Perhaps it is unrealistic to have a standard in which no instances of inauthenticity are ever allowed to exist.

It is helpful to return to the national security scenario one more time. In his work on the *infosphere*, ethicist Luciano Floridi suggests that it is essential to keep this area pristine and free of various types of pollution – including hate speech, fake news, propaganda, and perhaps counterfeit products like artificially produced voices and deep fake videos. In his mind, the producer of inauthentic materials contributes to the pollution of the infosphere. In his work, he also describes a future in which we will live in a "mature information society" in which we will have the same trust in the information running through our internet systems as we do for the water which comes through our pipes. Just as we expect in the developing world to turn on our water taps and receive fresh, drinkable water, he suggests that ideally, we will expect to turn on our information taps and receive reliable news and information. In this way, the person who conveys false or inauthentic information is seen as an obstacle to creating a *mature information society* (Floridi, 2013).

Psychologists might also point to the problems of growing up in an environment surrounded by authenticity or fake creations. It is terrible for people's mental health if they continually compare themselves to standards of beauty that are unattainable because the images they view have been manipulated and airbrushed. Inauthenticity can thus prevent people from enjoying the eudaemonia or human flourishing that Aristotle describes since they cannot fully enjoy being their authentic selves. It may also lead to issues like eating disorders or people spending too much money trying to achieve unattainable standards in terms of material goods.

Finally, legal scholars have suggested that creations like deep fakes present ethical issues as their creation may represent a theft of someone's original work or an intellectual property violation. At stake here is a question of ownership. Who could be said to own the image of Bourdain or the image of him reading from his diary?

Guignon (2008, p. 278) writes that "to say that a painting is an authentic Rubens is to say that it was (largely) painted by Rubens, that it came from his hand, in contrast to forgeries, fakes, imitations, mechanical reproductions, and so forth."

The World of Artificial Intelligence and the Ethics of Simulation

Here a more significant issue is whether anything produced by artificial intelligence (such as a novel or a poem) can ever be viewed as "authentic." Are robots inherently servile and inauthentic compared to humans? Can something produced by a program or an algorithm ever be superior to something produced by a human?

Some ethicists worry that artificially intelligent agents' artificial, simulated environments will lead to an overall devaluing and lessening of authenticity in modern life. Kraemer (2011) argues that "artificial means will always lead to an inauthentic result":

> watching a video of someone performing an activity is less satisfying than performing it yourself; socializing remotely is a poor substitute for socializing in person, and artificial or technologically mediated actions are the basis for an artificial rather than authentic life.

In his work, Kraemer (2011) describes a future in which people may have a neural interface that can cause them to feel happy, for example, when they are not. In his view, these people are being subjected to ethical harm even if they do not directly suffer ill effects from this manipulation. (That is, he argues that it would undoubtedly be ethically wrong to manipulate people's perceptions of risk to send them into a harmful battlefield simulation – but it would be equally harmful to cause them to believe that they were happy when they were not.)

However, this view is a subject of contention. A utilitarian might again suggest that avoiding the harmful effects of post-traumatic stress disorder or recovering from grief would be a sufficient reason to allow people's emotions to be manipulated through a neural interface. In this view, neural interface medical interventions would be no different ethically from a pharmacological intervention like prescribing anti-depressant medication.

Artificial Intelligence and the Problem of Agency

Throughout this book, we have encountered situations where the artificially intelligent bot could move beyond a status controlled by or answerable to a human being to a new status where the AI itself appeared to be in control. Through developments like machine learning and neural nets, AI programs can now perform actions like repairing their code or deriving new principles, including decision-making based on what they already know about a subject. Indeed, much of our concern about the transparency of algorithms comes from precisely this fact – that artificially intelligent programs are now often capable of deciding for themselves, and that, as mere humans, we may sometimes not wholly understand the mechanisms by which AI makes decisions – making it difficult for us to judge the accuracy of those decisions or even sometimes to undo them.

And, as we have seen throughout this text, the application of algorithms and automated decisions can have far-reaching consequences for individuals in all areas of their lives. Being denied credit or a job or being injured through a machine or algorithm's actions are often consequences that cannot be undone. Instead, the stigma of being misidentified as a terrorist or physical, mental, and bodily harm from having a drone fire upon one's business or home cannot be undone.

For that reason, designers today often accept that situations may differ and that some kinds of decisions and actions can be appropriately carried out by an autonomous system (such as perhaps killing vermin or bugs in your yard, deciding what type of fertilizer to apply to crops, or repairing small appliances). In contrast, other decisions with life-altering consequences might require a human to sign off or make decisions. (For example, using AI to grade papers, including standardized tests, may affect people's university prospects.)

Box 7.3 Critical Issues: Should Artificial Intelligence Be Allowed to Hack?

Today, many organizations and companies that rely on technology incorporate artificial intelligence into their decision-making processes. Artificial intelligence-powered algorithms are ubiquitous and incorporated into a variety of activities that people perform every day. For example, AI-powered voice recognition can help refer you to the correct department when you phone a company, and AI technology can suggest new products when you shop on Amazon.

Adding AI to a system no longer requires a high degree of technological sophistication for those who design systems. Instead, off-the-shelf programs can add artificial intelligence to one's business or production activities. Advertising firms, ticketing agencies, and even political consulting firms increasingly find that they rely on systems like IBM's Watson Studio or Salesforce's Einstein to perform various functions. AI-enabled systems can analyze your data, perform maintenance and enhancements, and even run experiments on your data before choosing a solution to an identified problem.

But, as Bruce Schneier (2021) warns, we face a solution where soon artificial intelligence programs may not merely be a component that goes along for the ride as we perform our everyday tasks. Instead, we may find ourselves in a situation where AI is in the driver's seat, making decisions on our behalf, often without our knowledge or consent.

Schneier argues that hacking encompasses a wide range of activities but that, in essence, "hacking" is merely a problem-solving method – which often relies on finding a backdoor or a loophole in the rules governing how a system works. And, since AI systems are often designed to find the most efficient solution to a problem, AI-enabled algorithms and bots may often end up hacking a system by applying the rules in a new and novel way that is often different from what a system's designers may have intended, or anticipated (Simonite, 2018).

Here a brief example can illustrate what this sort of hacking might produce and the consequences. When this author's children were young, they enjoyed playing a computer game called Zoo Tycoon. The game's goal was to establish a good zoo, and the indicator of how well one was doing was the amount of money the zoo was earning. But one could argue that a secondary goal of such a game was for young children to enjoy creating a beautiful environment and enjoy taking care of the animals in the zoo. My children figured out that if an animal in an enclosure fell ill, it was more efficient to feed the sick animal to the other zoo animals than to pay for a veterinarian visit and purchase additional food for the healthy animals. Thus, they earned additional points and ostensibly "won" the game by mistreating both the sick and the healthy animals. The solution they arrived at was correct and efficient, but it also violated the spirit of the game and arguably constituted an ethical breach. (In this way, it functioned similarly to a cheat code that one might find online while playing The Sims or another online game. Some players may find that using cheat codes does not personally feel like an ethical breach to them, while others might disagree.)

An AI is likely to arrive at a similarly efficient but morally ambiguous solution – because it lacks the awareness of the "spirit of the game" or the context in which a decision is being made. Instead, it will merely seek the most efficient solution. (For example, rather than fixing a broken sensor, it might choose to disable it so that it no longer indicates something is wrong.)

As Schneier points out, AI systems may even hack one another in the future, with little regard for how humans might constitute collateral damage. Here, we might consider today's dueling technologies available in plagiarism and anti-plagiarism software programs. High school and college students can access a site like Quillbot, which will paraphrase any text they enter, including a plagiarized term paper.

Quillbot's algorithm is designed to outwit anti-plagiarism software programs like Turnitin or SafeAssign, which search the web for matching text to identify when the material has been plagiarized. Thus, we might argue that Quillbot and SafeAssign are engaged in a competitive hackathon, as Quillbot aims to create a way of plagiarizing that SafeAssign can't detect. In contrast, SafeAssign aims to catch Quillbot in the act by increasing the sophistication of its plagiarism detecting algorithms. Neither Quillbot nor SafeAssign actually "cares" whether or not the students are learning anything in their college courses, and neither program's actions appear to consider students in the equation.

As such programs become more sophisticated, it is increasingly likely that human operators will be unable to fully comprehend the sorts of calculations that the algorithms are using, nor will they be able to undo them or modify them easily. Sometimes, we may know that an algorithm works, performing the assigned task, but may not understand how the program arrived at its conclusion.

In the future then, it will be essential to consider whether regulations can place limits on the ability of artificial intelligence programs to hack their own or other's programs, and how we can ensure that in seeking efficient solutions, the systems also mirror the values of their designers, rather than departing from them.

Bibliography

Schneier, B. (2021). *The coming AI hackers*. Cambridge, MA: Belfer Center, Harvard University.

Simonite, T. (2018, August 8). When bots teach themselves to cheat. *Wired Magazine*. Retrieved June 2, 2022, from www.wired.com/story/when-bots-teach-themselves-to-cheat/.

In her work, Amoore (2021) argues that today artificially intelligent programs are not merely the "clients" of human developers but agents acting to shape our world. She argues that in deciding how to organize data, what labels to attach to data, and how to describe relationships between data, robots make our social world and ultimately affect how we as humans who interact with machines will understand that world. She gives an example of a facial recognition program that might break down faces into different categories. While humans might be particularly concerned about breaking down racial barriers and not distinguishing between individuals based on race or disability, an AI program might make such issues more salient by classifying people based on these characteristics. She writes, "these algorithms may define what is normal behavior and what departs from normal parameters." For example, algorithms can distinguish between normal and deviant behavior. In this way, AI can be seen as helping to perpetuate injustices that humans may have committed to confronting and eliminating.

For this reason, in recent years, policymakers and software designers have begun articulating principles for how humans and machines should interact and the levels of control delegated to AI.

Meaningful Human Control

Meaningful human control (*MHC*) refers to design principles allowing responsible innovation. MHC principles are often applied when an ethical breach is possible, such as designing autonomous weapons. MHC establishes two conditions for system designers: tracking and tracing. Tracking means that the moral reasoning of the designers and the environment should be built into the system. MHC does not allow the machine to derive its moral calculus through artificial intelligence and machine learning. Instead, specific rules would be given to the machine by engineers (for example, a prohibition on killing civilians and mechanisms by which a system could identify civilians through techniques like facial recognition and environmental analysis). Tracing means that it is always possible to trace an action back to a human in an adverse outcome (de Sio & van den Hoven, 2018).

However, Ekelhof (n.d.) argues that MHC's focus is too narrow since it emphasizes the relationship between the operator and the machine when deployed. She asked for a broader rendering of the relationship between man and machine and all faces, including those leading up to the deployment of weapons. Here she describes the dynamic targeting cycle, which includes six phases – find, fix, track, target, engage, and assess. It is not enough, she argues, to allow a human to give the "go command" or push a button at the final phase. Instead, she argues that humans need to engage in better oversight overtargeting.

Making Ethical AI Policy: Who Decides?

As noted throughout this text, different cultures may sometimes have slightly different orientations towards particular ethical problems, even when they share similar values and value commitments.

Today, many nations are working to articulate a national or regional set of standards for adopting ethical artificial intelligence. Australia has developed an innovation strategy, and Denmark has adopted a strategy for digital growth (Radu, 2021). The US military and the North Atlantic Treaty Organization (NATO) have begun to speak about digital transformation.

In many instances, states have articulated a particular threat that they envision from how artificial intelligence might develop without a strategy or set of steering principles ethically in society. France warns against a winner-take-all approach by dominant actors in which one leading company may end up defining policies for how artificial intelligence is applied overall – through setting policies regarding questions like intellectual property, disinformation, how hate speech will be handled, and how data will be tracked and stored.

Indeed, even in repressive nations like Russia, a great deal of the innovation in artificial intelligence has come from private actors and private corporations. Private corporations do not always have the same commitment to ethical behavior or the same desire to regulate ethics as a government entity might have. Computer science practitioners will likely play a key role in such a situation, as articulators and designers of ethical artificial intelligence policies and procedures and perhaps as industry leaders.

Ethical Responsibilities for AI Professionals

In this chapter, we have discussed ethical principles related to responsibility. We have asked how humans who develop and administer AI programs should and should not be held responsible for the actions an AI commits. Ethicists have long distinguished between levels of responsibility when assessing wrongdoing. Barry et al. (2014) distinguish between enabling, doing, and allowing harm. One who commits an act directly harming is viewed as morally and ethically responsible for the act. One who is aware of wrongdoing (such as a coworker participating in embezzlement) but does not do anything to stop it may be viewed as allowing the harm since they did not act to prevent it. Finally, someone who served as a lookout while the coworker was using a computer to alter financial records to commit fraud might be viewed as enabling harm. The causal relationship is different in each case, ranging from acts explicitly caused to those that merely occur with one's knowledge.

However, ethicists differ in the degree to which people may be viewed as complicit in a situation where they do not commit an immoral act but instead enable or allow harm. Manjikian (2015) describes a situation that occurred in the Netherlands in 2014. American intelligence agents collected information from Dutch communications satellites, ostensibly tracking terrorist activities in Western Europe. However, the information was subsequently used to conduct deadly drone strikes on terrorist camps in Somalia. Dutch citizens were outraged when they learned about these events since public opinion overwhelmingly opposes using drones in war. The ethical issue presented is this: Were the Dutch authorities somehow responsible for anticipating how the information they shared with the Americans would be used, and were they complicit in any subsequent deaths which occurred since they had in some ways participated in or supported the act that occurred?

As they think about their professional responsibilities, individuals working in the field of artificial intelligence have begun asking similar questions. Is it enough just to do good work? Or does the moral end to which such work leads matter? As Bellanova et al. (2021) remind us, computers and AI are potent technologies. They can be used as force multipliers to enhance our abilities to carry out actions. An algorithm could be used to distribute food to people after a natural disaster more efficiently, or it could be used to increase the efficiency of police forces in cracking down on demonstrators through facial recognition software. AI technologies could make military targeting more accurate and enable us as humans to act in more deadly and violent ways.

Thus, practitioners in this field ask questions like: Does building an artificial intelligence program that can be used for mass surveillance, for example, count as enabling harm? And do practitioners have an ethical duty to consider the aims to which their work will be put and to refrain from participating in projects that might lead to anti-democratic or unethical ends?

In 2014, the Future of Life Institute was founded to consider ethical questions related to emerging technologies. In 2018, the first draft of the Asilomar Declaration on the ethics of artificial intelligence was put forward. The precedent for such declarations of principles exists. Biology, chemistry, and nuclear physics researchers have asked similar questions about their professions. They have vowed to consider the moral import of producing goods like biological, chemical, and nuclear weapons. In each case, researchers have promised to carry out research in an ethically responsible manner, considering the ends to which their knowledge might be put.

The *Asilomar Declaration* has been signed by nearly 4000 researchers working in artificial intelligence and related fields. It contains 23 principles, covering such categories as research, ethics and values, and long-term issues. This text includes many values: safety, transparency, human control, privacy, and responsibility. It also includes the goals of avoiding an artificial intelligence arms race, avoiding risks, and working for the common good (Crowe, 2017).

While it is essential that such principles be codified into government laws and regulations, it is equally essential for practitioners to publicly subscribe to such principles and commit to upholding them as research goes forward in this challenging and unpredictable field. The Morningside Group is a conference of professionals representing international organizations like Google and Kernel and research scientists in several countries, including Canada, Israel, China, Japan, the United States, and Australia. This group has come together to discuss and articulate ethical principles specifically in artificial intelligence and neural implants. They identify ethical issues in privacy and consent, agency and identity, bias, and augmentation or enhancement. These members also recommend that international treaties like the 1948 Universal Declaration of Human Rights be amended to codify more specifically the rights and responsibilities of individuals in the areas of working with artificial intelligence. They also advocate for creating an international convention (like the UN Conventions on Biological and Chemical Warfare or Landmines) that would define prohibited actions related to neurotechnology and machine intelligence. In addition, the *IEEE*, the world's largest organization of technology professionals, began a global ethics initiative in April 2016 to embed ethics into the design of processes for all AI and autonomous systems.

Those who will work with data and with artificial intelligence are also urged to think preemptively – considering not only what sorts of ethical issues their work raises today but also what sorts of ethical issues might arise in the future, given the trajectory and evolution of these rapidly changing and powerful technologies.

CHAPTER SUMMARY

- Ethical challenges related to artificial intelligence and its applications include the challenge of authenticity, or the fact that a simulated experience may not be the same as a genuine experience.
- A second ethical challenge relates to agency – or the ways in which humans can cede decision-making authority to algorithms and nonhuman decision-makers.
- Autonomy describes the ability of an individual or entity to act freely and unimpeded, making its own decisions and choosing to carry out its own actions.
- At present, autonomy granted to machines occurs on a spectrum, from partially to fully autonomous.
- Humans can take the role of a decider when working with machines through a human in the loop and meaningful human control (MHC) design principles.

DISCUSSION QUESTIONS

1 Artificial Intelligence and Search Algorithms

Visit Google's pages to explore the materials they have put together about their organization. In particular, considering watching the documentary about Google Search, its algorithms, and planned changes to those algorithms.
Available at google.com/search/howsearchworks
This one-hour-long movie includes several ethical issues:

- How a search engine should treat materials that might be promoting a hoax (though considering notions of relevance and authoritativeness)
- How "autocomplete" sometimes returns offensive suggestions (How does Google flag "inappropriate predictions"?)

For discussion:

- What are some ETHICAL issues that Google raises in the first half of this documentary?
- What did you learn about Google's policies regarding hate speech and accountability?
- What are some challenges that Google is facing in the areas of machine learning and Google Search (i.e., BERT in search; Deep rank)?
- Does the increased use of machine learning in Google Search remove humans from the "driver's seat"? Should humans have more agency in deciding which materials they use online and how they are selected, or is this simply a technological impossibility given the available materials?

2 **Ethics and Authenticity**

In a notable book written in 1972 called *What Computers Can't Do*, Hubert Dreyfus reflected on the idea that it would someday be possible to create a computer that emulated what humans do. He described the notion of an "effective procedure," or a "set of rules which tells us, from moment to moment, precisely how to behave" (p. 67). However, Dreyfus argued that being human means having a physical body that moves in and interacts with the world. He argued that many very optimistic people about AI technology fail to understand the "fundamentally different nature of man and machine" (Dreyfus, 1972).

Watch the following clip of three computers designed by Boston Dynamics dancing to the song "Do You Love Me?" Available at www.youtube.com/watch?v=fn3kwm1kuaw (accessed June 21, 2022).

- Do you think it is appropriate to say that these computers are "dancing"?
- What would be the differences between what humans do when they dance and what robots do when they dance?
- How does this clip show the possibilities and limitations of what robots can do compared to what humans can do (and computers can't do)?

3 **Ethics of Brain-Computer Interfaces**

Watch this eight-minute YouTube presentation about computer entrepreneur Elon Musk's Neuralink brain-computer interface (BCI) device. Elon Musk. January 9, 2021. "We Won't Need Language." Available at https://youtu.be/R6ok4NI4X8c (accessed June 21, 2022).

Imagine you could have Neuralink implanted and have your brain wired together with several others in your work unit. This could make your work unit more efficient and allow you to share more data without explicitly writing or speaking it to your partners. You may not have to use language altogether!

- Would you take the opportunity to join this program? Why or why not?
- What would be the pros and cons of doing so?
- What sorts of safeguards should be put into place to safeguard the individual rights of those participating in this "mind meld"?

4 **Culture, Value, and Ethics**

Virtues can be individual or social. Social virtues include cooperation and reliability in the sense of duty. Some philosophers believe that social virtues make sense only within a society that values those things.

Reflect upon the following question:

> As a society in the United States, do we value authenticity? Yes or no? If so, in what way, under what circumstances?

RECOMMENDED RESOURCES

Borgesius, F. Z. (2018). *Discrimination, artificial intelligence, and algorithmic decision-making*. Strasbourg: Council of Europe. Retrieved June 1, 2022, from www.ivir.nl/study-about-discrimination-and-artificial-intelligence-by-Frederik-Zuiderveen_Borgesius.

CNBC News. (2020, January 10). *The hype over quantum computers, explained*. Retrieved June 21, 2022, from www.youtube.com/watch?v=u1XXjWr5frE.

de Sio, F., & van den Hoven, J. (2018). Meaningful human control over autonomous systems: A philosophical account. *Frontiers in Robotics and AI*. Retrieved June 21, 2022, from https://pubmed.ncbi.nln.nih.gov/33500902.

Future of Life Institute. (2017). *Asilomar artificial intelligence principles*. Retrieved June 1, 2022, from: https://futureoflife.org/2017/08/11/ai-principles/.

Thomas, R. *A website with eleven short videos about Machine Ethics and AI*. Retrieved June 1, 2022, from www.fast.ai/2021/08/16/eleven-videos/.

References

Ahronheim, A. (2021, May 27). Israel's occupation against Hamas was the world's first AI war. *Jerusalem Post*. https://m.jpost.com/arab-israeli-conflict/gaza-news/guardian-of-the-walls-the-first-ai-war-669371/amp?__twitter_impression=true.

Amoore, L. (2021). Deep borders. *Political Geography*. https://doi.org/10.1016/j.polgeo.2021.102547.

Barry, C., Lindauer, M., & Overland, G. (2014). Doing, allowing and enabling harm: An empirical investigation. In J. Knobe, T. Lombrozo, & S. Nichols (Eds.), *Oxford studies in experimental philosophy: Volume 1*. Oxford: Oxford University Press.

Bellanova, R., Irion, K., Jacobsen, K., Ragazzi, F., Saugmann, R., & Suchman, L. (2021). Toward a critique of algorithmic violence. *International Political Sociology*, *15*(1), 121–150.

Blackford, R. (2012). Robots and realities: A reply to Robert Sparrow. *Ethics and Information Technology*, *14*(1), 41–51.

Crowe, S. (2017, February 3). Asilomar AI principles: 23 Tips for making AI safe. *Robotics Business Review*. www.newscientist.com/article/2143499-ships-fooled-in-gps-spoofing-attack-suggest-russian-cyberweapon/.

de Sio, F., & van den Hoven, J. (2018). Meaningful human control over autonomous systems: A philosophical acount. *Frontiers in Robotics and Artiicial Intelligence*, *5*(15). Retrieved June 2, 2022, from www.frontiersin.org/articles/10.3389/frobt.2018.00015/full.

Dreyfus, H. (1972). *What computers can't do: A critique of artificial reason*. Boston, MA: MIT Press.

Ekelhof, M. (n.d.). Autonomous weapons: Operationalizing meaningful human control. *Law and Policy Blog*. http://blogs.icrc.org/.

Feldstein, S. (2019). The road to digital unfreedom: How artificial intelligence is reshaping repression. *Journal of Democracy*, *30*(1), 40–52.

Floridi, L. (2013). *The ethics of information*. Oxford: Oxford University Press.

Gent, E. (2020). Taming the bots. *New Scientist*. Retrieved June 3, 2022, from www.sciencedirect.com/science/article/abs/pii/S0262407920322454.

Guignon, C. (2008). Authenticity. *Philosophy Compass*, *3*(2), 277–290.

Hambling, D. (2017, August 10). Ships fooled in GPS spoofing attack suggest Russian cyberweapon. *New Scientist*. www.newscientist.com/article/2143499-ships-fooled-in-gps-spoofing-attack-suggest-russian-cyberweapon/.

Hardin, R. (2006). *Trust*. New York: Polity.

Henderson, K. E.-K. (2019). "Oops, I did it" or "It wasn't me": An examination of psychological contract breach repair. *Journal of Business and Psychology, 35*(1), 347–362. https://doi.org/10.1007/s10869-019-09624-z.

Karami, A., Lundy, M., Webb, F., Turner-McGrievy, G., McKeever, B., & McKeever, R. (2021). Identify and analyzing health-related themes in disinformation shared by conservative and liberal Russian trolls on Twitter. *International Journal of Environmental Research and Public Health, 18*(4), 2159. https://doi.org/10.3390/ijerph18042159, PMID: 33672122; PMCID: PMC7927016.

Kraemer, F. (2011). Authenticity, anyone? The enhancement of emotions vai neuropharmacology. *Neuroethics, 4*(1), 51–64.

Ma, Y., Xie, T., Li, J., & Maciejewski, R. (2019). Explaining vulnerabilities to adversarial machine learning through visual analytics. *IEEE VAST (Transactions on Visualization and Computer Graphics). IEEE Conference Proceedings*. Retrieved from https://ieeexplore.ieee.org/xpl/conhome/8960971/proceedings.

Manjikian, M. (2015). But my hands are clean: The ethics of complicity and the problem of intelligence sharing. *International Journal of Intelligence and Counterintelligence, 28*(4), 692–709.

Petrella, S., Miller, C., & Cooper, B. (2021). *Russia's artificial intelligence strategy: The role of state-owned firms*. Philadelphia, PA: Foreign Policy Research Institute.

Radu, R. (2021). Steering the governance of AI: National strategies in perspective. *Policy and Society, 40*(2), 178–193.

Royakkers, L., Timmer, J., Koo, L., & van Est, R. (2018). Societal and ethical issues of digitization. *Ethics and Information Technology, 20*(2), 127–142.

Schlosser, M. (2019). Agency. In E. Zalta (Ed.), *The Stanford encyclopedia of philosophy, winter 2019 edition*. Retrieved from https://m.jpost.com/arab-israeli-conflict/gaza-news/guardian-of-the-walls-the-first-ai-war-669371/amp?__twitter_impression=true.

Shorey, S. (2016, February 24). What is it like to be a bot? *Data and Society: Points*. Retrieved from https://philosophynow.org/issues/126/What_is_it_like_to_be_a_bot.

Sparkes, M. (2021, April 24). Will the EU save us from AI dystopia? *New Scientist*. Retrieved June 1, 2022, from www.sciencedirect.com/science/article/abs/pii/S0262407921006837.

Taddeo, M., & Floridi, L. (2017). Regulate artificial intelligence to avert cyber arms race. *Nature*, 18–20.

Tangermann, V. (2017, June 2). Researcher saysn an image generating AI invented its own language. *Futurism*. https://futurism.com/researcher-image-generating-ai-invented-language.

8 Cybersecurity, Diversity, Equity, and Inclusion

LEARNING OBJECTIVES

At the end of this chapter, students will be able to:

1. Define algorithmic bias and give examples
2. Describe how machine learning depends on datasets that can reflect historic injustices
3. Define issues related to equity and inclusion – and describe how data can be both examined for attention to these issues and improved in the future
4. Describe existing specific commercial and state-led policy initiatives aimed at combatting these types of injustices in cyberspace

Today, more than ever, we depend on technology in all areas of our lives – from education to travel to employment and shopping. But how attuned are the technological systems that rely upon differences in how they work and work differently for those of different genders, races, and national origin groups? Here, we can consider the following five examples:

- In 2018, the American Civil Liberties Union tested Amazon's facial recognition program. They used photos of US House of Representatives members and the US Senate. Many of the photos were matched to existing photos in databases incorrectly, and in some cases, members of Congress were "identified" as criminals whose faces appeared in a criminal database. Those who were misidentified were much more likely to be African American or Hispanic. Facial recognition systems also struggle to identify women and people with darker skin (Wong, 2019).
- In 2019, Britain's Home Office (the government department responsible for domestic affairs in the United Kingdom) deployed a face-detection system for its passport photo-checking service despite being aware that it didn't work well for those with very light and dark skin. The office worked with a vendor to create a fix, but it hadn't been applied more than a year later. Some Black users of the passport service were told their photos didn't meet the requirements. Asian users were sometimes told their photos were unacceptable because they had their eyes closed even when they did not (Vaughan, 2021).

And how good is your government at identifying potential terrorists? Nick Wing describes the information that could be used to add your name to the *US Terrorist Identities Datamart Environment* (*TIDE*) and then into the *TSDB* (*Terrorism Suspects Database*). If

DOI:10.4324/9781003248828-11

your name appears in the TSDB, you can be prevented from taking flights in the United States, investigated further by law enforcement authorities, or even denied certain types of employment, such as jobs requiring a security clearance. One might be nominated as a potential terrorist based on who you associate with online, the language and phrases you use, articles online that you might have "liked," and whether someone you are related to appears on the list (Wing, 2014).

- In 2021, a US health task force began studying the algorithm to assign places to people on the waiting list for a kidney transplant in response to allegations that the algorithm contributed to race-based health inequities. Critics stated that the algorithm asked a question about race that routinely classified Black patients as healthier than other patients with the same blood tests. As a result, Black patients waited longer to receive a kidney transplant and had worse outcomes as they were often sicker when a kidney became available (Bichell & Anthony, 2021).
- In 2020, the British government decided to cancel A-levels, the standardized exams students take at the end of their secondary school careers, due to COVID risks. Students apply to university in their final year of college based on their school performance but are not formally admitted until they submit A-level marks. Students with bad A-level marks may thus see their admissions revoked. In 2020, Britain's Education Secretary worked with a commercial firm named Ofqual to create another measure of academic achievement without A-levels. The algorithm applied to students' grades was meant to "standardize" marks so that they could be compared across the schools in Britain. Ofqual used a measure that depended on schools' past performance and other factors. As a result, 40 percent of students saw their grade point averages downgraded. Students from deprived areas were disproportionately affected and worried that these changes would affect their university prospects.
- In the United States, the Department of Homeland Security routinely uses *ATLAS software* to mine federal databases, looking for derogatory or harmful information about immigrants to the United States. Information from immigrants' visa applications is routinely entered into various federal databases, including the FBI's Terrorist Screening Database and the National Crime Information Center database. The ATLAS program looks for evidence of fraud and dishonesty and whether someone is potentially dangerous. In 2019, ATLAS conducted over 16 million screenings and flagged over 100,000 cases as potentially suspicious. Some groups have spoken out against this practice, pointing to flaws and inaccuracies in the databases used for screening. A US Muslim group (Muslim Advocates) has spoken out against the software, noting that its purpose is to deny citizenship rather than offer it. They have also asked for greater transparency regarding how these decisions are made, accusing the US Department of Homeland Security of operating according to a byzantine set of "secret rules" – including predictive analytics and algorithms. Immigration advocates have criticized Amazon for allowing this program to run on its cloud (Biddle & Saleh, 2021). Biddle and Saleh (2021) worry that "ATLAS is making potentially life-ruining decisions based on bad data." Indeed, the United Kingdom has already abandoned its algorithmic system for granting immigrants visas. Immigration advocates in the United Kingdom have also warned Britain not to share data with the United States, criticizing agreements like the data-sharing agreement between the National Health Service (NHS) and Palantir, a commercial firm used by US Immigration and Customs Enforcement (ICE).

These stories all illustrate some key takeaways when we begin to think about diversity, equity, and inclusion in the context of cybersecurity. First, these stories should be cautionary tales, illustrating how algorithms are not perfect. Machine learning experts sometimes use the phrase "garbage in, garbage out" to illustrate that artificial intelligence needs to be trained on a dataset and that the AI's accuracy thus depends on the dataset itself. Suppose a dataset contains partial data (i.e., not including sufficient data about women, ethnic minorities, or sexual minorities). In that case, the resulting insights will reproduce these biases rather than correct them.

In addition, as Debrusk (2018) notes, "machine learning models are, at their core, predictive engines. Large datasets train machine learning models to predict the future based on the past." As a result, we can imagine a scenario in which a machine learning algorithm is trained to know about particular professions. As part of that training, the machine learning algorithm is fed large amounts of data, including images taken from Google Images. As a result, the algorithm "learns" that men are doctors and women are nurses, since the vast majority of doctors (at least in the United States) have been men and the majority of nurses have been women. While this fact is true historically, it does not reflect today's reality, nor does it reflect most people's aspirations to create a more fair and equitable world in which anyone who wishes to become a doctor and has the necessary qualifications can become one. Machine learning models can thus often reflect and ultimately reproduce social inequities, not as the result of conscious choices but rather as the result of historical data used to train a machine (Debrusk, 2018).

Next, these stories illustrate the differential impact of insufficient data, inadequate training, and bad algorithms developed and applied. In each story, the damage or risk resulting from the flawed algorithm was not equally distributed across society. Instead, those who might have the fewest resources to organize and dispute a finding made by an AI are often subject to bad decisions. Someone who is denied a loan due to their immigration status, education status, English-language-learner status, or gender is perhaps less well-equipped to dispute the decision than someone wealthy, well-educated, and a majority group member.

And these stories all illustrate how a project that might have begun in a laboratory, often due to intellectual interest, makes the leap from that laboratory environment into the real world. It imposes real-world consequences on those affected by an AI's decisions. Being misidentified as a potential criminal or a potential terrorist may have severe and long-lasting economic, political, and social consequences for an individual or a family. The power of algorithms is apparent. Thus, it is essential to ensure that they are accurate so that racial, socioeconomic, gender, and other inequities are addressed rather than multiplied when algorithms are trained, applied, and used.

Furthermore, these stories show how algorithms' decisions are often opaque or incapable of being identified and understood by those subjected. In the outcry about admissions to British universities, students and parents demanded to know more about how the algorithm worked and to be allowed to have input into critiquing the algorithm and its application. However, in many instances, people do not even know that a decision has been made using an algorithm, nor are they informed about how it has been constructed.

These stories also show how biased algorithms and insufficient data exacerbate social inequities. Those who were most likely to be denied a place at university due to the A-level algorithm in the United Kingdom were students from the poorest schools who had already faced the most significant challenges in securing an education. The harsh consequences of lousy decision-making fall disproportionately on those already operating from an unequal position.

Finally, such stories illustrate that we most likely live in a technologically mediated world, no matter who we are or what we do today. Technology affects how we see and experience our world and live in it. It affects who else we come into contact with, how we interact with these people, and how we see them and ourselves.

Therefore, it is essential to consider how people's lives are lived within a technologically mediated environment and whether everyone experiences the world the same way in such an environment. Are some groups of people more or less comfortable living with technology? Is the power that technology conveys applied in the same way towards everyone? In other words, the availability of technology may make the world feel safer for some groups of people while making it feel more dangerous to others. Do some groups find their ideas and identities validated online in ways others do not? Do all groups experience surveillance in the same way? Does everyone worry to the same degree about online privacy threats? Who is more likely to be stalked or harassed?

How Governments Use Data

Governments have always used data to "see" their subjects, know who lives within a state's borders, and provide for and plan for subjects' needs more effectively. The taking of a census of citizens has been around since the Roman Empire, several thousand years ago. Censuses were used to plan for national disasters like an agricultural shortfall and ensure a sufficient workforce was available for a nation's army and navy.

However, as noted in Chapter 5, governments can now collect and use data in novel and unprecedented ways. And using algorithms and Big Data, states can better make predictions about future needs and citizens' activities – identifying some as posing a potential security risk or at risk of harming themselves or others.

On the surface, one can identify many benevolent reasons why a nation might want to collect data about its citizens – as noted earlier. Indeed, many philosophers of technology argue that data itself is ethically neutral. It is not inherently good or bad. Instead, the difference rests on what those in authority choose to do with the data they collect.

In *IBM and the Holocaust: The Strategic Alliance Between Nazi Germany and America's Most Powerful Corporation*, journalist Edwin Black illustrates this principle. In his work, he describes how the US corporation International Business Machines (IBM) worked with Nazi Germany throughout World War II, furnishing the Nazi regime with a primitive computer-based technology based on punch cards which allowed them to more precisely identify and target Jewish families, Roma families, and sexual minorities for extermination. Using what was then a "high-tech" solution, the Nazi regime was able to assign prisoners with specific skills to specific work camps and calculate the minimum servings of food to sustain prisoners in a work camp or ghetto.

Black thus advises caution when thinking about new technologies and the harms they might cause. In his work, he makes the moral and ethical argument that IBM and its executives were complicit and morally guilty for their part in helping to uphold and support the Nazi regime through furnishing and servicing its technology. He argues that they knew how their technology was being used and should have acted to prevent it from being used this way.

His history also illustrates how many ethical values – such as efficiency or the ability to quickly achieve a task with minimal resources – may seem virtuous or morally upright in the abstract while being morally wrong within specific contexts. The same data collection and analysis principles that help planners make efficient and equitable decisions can also be used for other morally ambiguous ends.

For this reason, it is essential that those working in cybersecurity be capable of analyzing and critiquing emerging technological fixes to identify both the potential benefits and potential harms that can emerge from such technology. Using deontological ethics, we can also emphasize considering aggregate risks that an entire population might encounter in applying new technology and identifying specific risks and harms that might fall differentially on a particular group. It is essential to consider how those who are least powerful in our societies may be affected by applying algorithms and the compilation and use of Big Data.

Box 8.1 Tech Talk: Open Data Initiatives and Programs

Do citizens have the right to access and work with data collected by their governments? More and more, there is a consensus that they do.

Since 2015, many groups in many nations have begun pressuring government entities at all levels, including city-state and federal, to make as much information as possible freely available to all citizens, not just to view but to explore and work with online. Citizens have begun exploring databases showing their city's budgets, databases of government employees' salaries, and databases showing how the government allocates contracts for goods and services.

Proponents of this approach argue that citizens have a right to this data since government funds carry out data collection activities. If my tax dollars fund the census, citizens argue, then by rights I should have access to this data.

Furthermore, citizens stress that open data serves the ethical goal of providing greater transparency and making the government more accountable to the citizens. Watchdog groups and citizens can examine budgets for evidence of political corruption, kickbacks, waste, fraud, and abuse. Citizens have explored government data to see if criminal sentencing differs depending on a person's race, gender, or immigration status. Citizens can also see how resources like education spending or policing resources might be allocated within their cities.

Currently, 30 governments worldwide are signatories to the *Open Data Charter*, which pledges to make data accessible to citizens. However, as the Open Data Barometer report shows, only about 20 percent of all government datasets are accessible to citizens. The United Kingdom is taking the lead in the open data initiative, assuming that the default setting for government data should be one of openness (World Wide Web Foundation, 2018).

Currently, those who support open data initiatives are concerned that new datasets be made available in machine-readable formats so that information can be easily cleaned, loaded, and transformed. The datasets can be easily queried and explored. In addition, they would like to see governments adopt procedures whereby new data is automatically loaded to download sites without delays caused by bureaucratic approval processes.

Conflicting Ethical Values: Openness Versus Security

However, some critics of the open data approach have raised concerns about the privacy of individuals and groups whose data might be publicly available. For example, in a smaller city, even if the salaries of government employees are only listed by rank or position, people might still be able to narrow down the data to identify specific individuals. These critics argue that the goal of making all data open and available may be too broad and that individuals should still be involved in making decisions about what sorts of data to release and what to keep private, rather than simply automatically releasing all data. (For example, statistics about disease prevalence or crime statistics that pose security concerns might still be kept private or classified.)

In addition, some privacy-producing tools described earlier in this text (such as the use of algorithms to alter and encrypt data) could not be applied to data that was also required to be made publicly available as part of an open data initiative. In this way, the two goals of openness and transparency versus security and privacy could conflict here.

Designing for Openness and Transparency

As part of the open data initiative, activists also call upon governments to think about openness and representation as goals in the planning and design of data and research activities, rather than adding on openness as a goal once data has been collected. Ideally, those who carry out research activities would consult with users and the data subjects, including those underrepresented groups like ethnic, sexual, and racial minorities, to ensure that the data collected represents a diversity of viewpoints and that questions and indicators are meaningful and sensitive to these concerns. They might also consider the formats in which data would be presented so that it was most useful to data subjects who might wish to work directly with the data themselves.

Bibliography

World Wide Web Foundation. (2018). *Open data barometer – leaders edition*. Washington, DC: Author.

What Is a Disparate Impact?

The stories which began this chapter showed how technology-based decision-making mechanisms often have a disparate impact on one group of individuals – consumers, medical patients, or employees – about all other individuals in the same group. In the American context, **disparate impact** refers to situations where a test or other tool is used for selection that may appear neutral but hurts a particular protected class of individuals. The *1964 Civil Rights Act* often provides the basis for court rulings involving disparate impact in the United States. That law forbids labor law practices that disproportionately hurt members of a protected class. In the United States, protected classes against which one cannot discriminate include the elderly, women, racial and sexual minorities, religious minorities, and those based on national origin (Gilliard & Golumbia, 2021).

In seeking to combat disparate impact, then, advocacy groups in the United Kingdom have stressed the importance of transparency. The UK group Foxglove Legal has emphasized the fact that those who are subject to disparate impacts (for example, in being denied a job or a bank loan) are often unaware that an algorithm has been used in arriving at that decision and are even less aware of how the algorithm might function. Furthermore, people may have been socialized to begin with, assuming that such decision-making mechanisms are fair. In this way, they are not likely to look for evidence of bias, or information about the sorts of bias that might exist in an algorithm.

For example, until recently, many automated hiring systems (such as a portal on LinkedIn where you upload your resume) used algorithms that ranked potential candidates by looking for characteristics they shared with those who already worked at the company. So if a company had a poor track record of hiring graduates of Historically Black College and Universities (HBCUs) or women's colleges, then it was unlikely to find itself interviewing members of these groups since the algorithm would have already marked them as being poor candidates for employment at that company. Automated hiring systems discriminate against racial minorities, women, and those who traditionally encounter age-based discrimination.

In addition, algorithmic decision-making may have a disparate impact on workers with disabilities (Konopasky, 2021). Researcher Aaron Konopasky calls our attention to how employers frequently use computerized sorting mechanisms and programs like online games to test whether an employee has the necessary skills for a position. He notes that prospective employees might be asked to take a personality profile or solve a problem to measure their traits and supply additional data to a computer algorithm. Tests might measure psychological traits – like whether an individual prefers to work alone or as part of a group and abilities like processing speed. Konopasky (2021) suggests that such tests might violate the Americans with Disabilities Act if they are used to screen out categories of employees. That is, they may be a form of discrimination. However, there is not currently a legislative act to combat this problem.

A recent article looking at online learning also described how a keyboard monitoring program called "Proctorio," used by many schools and universities to prevent cheating during tests and exams, might have a disparate impact on students. The reporters describe how the program requires students to hold up their laptops to give Proctorio a "tour" of their space to show no one else in the room helping the student. However, some students may not have a private space in their homes, and the requirement means that tests cannot be taken in a public space like a library. In addition, students may not have a quiet space, and the software may read noise from siblings or children as someone interfering with the test by rendering assistance. The requirement that students sit still and not move from their seats can also be detrimental for students with a disability that causes them to move or get up during the exam. Finally, some students have objected that the cameras work better for those with fair skin. A darker-skinned student might appear to have left the room if, for example, a shadow falls across the screen. In some instances (such as at the University of Illinois at Urbana-Champaign), students have successfully petitioned their university administration to stop using the software ("Proctorio's Awful Reviews Disappear Down the Memory Hole," 2021)

What Is Bias?

Disparate impact results from *bias*, defined as "systematic error introduced into a study sampling analysis by consciously or unconsciously choosing or promoting certain outcomes over others." As Dankwa-Mullan (2020) notes, "Bias implies error resulting in one group being favored over another." A biased outcome or ruling can thus come from an error such as relying on data that is itself biased (containing errors related to the data included or the labels applied to the data) or by relying on a decision-making algorithm that is flawed in some way in its calculations. In *Invisible Women: Data Bias in a World Designed for Men*, author Caroline Criado-Perez (2020) describes how medical trials of new medications and procedures routinely excluded women as test subjects in the United States for many years. Researchers argued that because women's health might differ throughout the month due to hormonal fluctuations or throughout their lifetime through events such as pregnancy and childbirth, the data generated by women as test subjects were simply "too messy" and imprecise to fit into existing models of health. As a result, Criado-Perez argues, women were significantly more likely to be misdiagnosed, given incorrect medications, and even die in auto accidents since "crash tests" were done using male bodies as the default car crash victim. Thus safety fixes in automobiles were designed to protect male bodies rather than female bodies (Criado-Perez, 2020).

Invisibility, or *data invisibility*, thus describes a situation where a particular group of individuals may not be "seen" within a dataset or algorithm. Data is often not collected about a group that suffers from data invisibility. Researchers may even have a "blind spot" that allows them to overlook this group's missing experiences and statistics. In her work, Nelson describes how, for example, contemporary higher education datasets often fail to consider and capture the experiences of Native American (indigenous) students who may attend American colleges and universities (Nelson, 2017).

Creating Data and Algorithms Which Are Fair and Equitable

The opposite of disparate impact is thus *fairness*. A fair algorithm is one in which each individual has an equal chance of receiving the same favorable or unfavorable rating regardless of their characteristics. As Reynolds (2017) notes, "fixing" an algorithm requires more than simply paying attention to how variables like race, socioeconomic status, or gender affect an outcome. Race, SES, or gender can also influence other variables, like where someone lives or their employment patterns. He writes, "it's not enough to ignore obvious race or gender cues; you have to consider how variables interact." The best way to do this is to make algorithms transparent – so that those who work with data and those subject to an algorithm's decisions can better understand how they work and identify any disparate impact cases due to algorithmic bias

It is important to note that discrimination based on disparate impacts may be unintentional. As noted earlier, it may result from a flawed algorithm or another technologically enabled decision-making algorithm. Nonetheless, at least in the European Union, a corporation or government entity can still be found guilty of *discrimination*, even if the disparate impact created was unintentional – due to an algorithm error or a designer's decision. Disparate impact can result from various factors, including defining target variables and class labels, how training data is labeled, how data is collected, and what other features are enabled in a machine learning program (Borgesius, 2018).

Today, many major platforms that host machine learning tools also include tools designed to check and correct for bias in models created. IBM's AI Fairness 360 is an open-source metrics toolkit designed to check for unwanted bias in datasets and machine learning models.

Meanwhile, Amazon is experimenting with a fairness measure called conditional demographic disparity. This measure, developed by the Oxford Internet Institute, asks,

> is the disadvantaged Class A a more significant proportion of the rejected outcomes than the proportion of accepted outcomes for the same class). In other words, is a group (like women or recent immigrants) being rejected by a program by a significantly higher proportion than the aggregate group or default group?

Here, it's important to note that while an AI can be taught to apply a metric like this, and perhaps even to adjust a model if bias is discovered, this is not the same as saying that "AI cares about fairness."

Amazon's SageMaker machine learning service also contains tools for identifying and correcting bias in algorithms. (Links to specific sites containing information on these tools are found under "Recommended Resources" at the end of this chapter.)

However, not everyone agrees that "de-biasing" the datasets that machine learning programs are fed is sufficient to create a just and equitable world or even one in which data is used equitably and fairly. In a report written for the European Digital Rights group, Agathe Balayn provides the following anecdote. She describes how in 2018, a query to Google Images for a photo of "three black teens" often returned images of mugshots from criminal databases, while a request for a photo of "three white teens" was more likely to return photos of teens going to the beach or engaging in a fun activity. In the report, she quotes the Senior Policy Advisor for the European Digital Rights group, Sarah Chandler, who says:

> concerning structural discrimination, whilst some of the harms to marginalized groups may relate to biases in data sets or system designs, the majority relate to how AI systems operate in a broader context of structural discrimination – including how they go about recreating and amplifying existing patterns of discrimination.
>
> (Balayn & Guises, n.d.)

In other words, if a machine learns that images of black teens include mugshots, this is not because the artificial intelligence is not acting correctly, but because our justice system acts incorrectly. Black males are more likely to be imprisoned. That is a structural problem, not a machine learning problem; therefore, she argues what is needed is not merely a technological fix but a more radical fix. Thus, Chandler critiques the European Union's emphasis on debiasing datasets and questions their utility, which she believes is overstated (Balayn & Guises, n.d.).

Increasing Transparency

Another concern for developers is the ability to increase users' and consumers' understanding of how algorithms created and administered by artificial intelligence work. *Algorithmic transparency* is defined as "openness about the purpose, structure and underlying actions of the algorithms used to search for, process and deliver information" (TechTarget Contributor, n.d.).

A related term is *explainability*. States that have regulations in place requiring that companies be able to respond to queries by consumers who wish to understand how a decision affecting them was made distinguish between two types of explanations. A global explanation is a complete explanation for the outcomes of a given process. Here, a consumer is unlikely to require a global explanation, which requires a description of the entire model used to make a decision – including all of the rules or formulas used and the relations among the input variables. A global explanation shows that fair procedures are being relied upon when deciding. A local explanation explains how your particular outcome was arrived at. Someone who wants to know why they (or people like them as a class) were denied a service like a mortgage would likely receive a local explanation (MacNish, Fernandezinguanzo, & Kirichenko, n.d.).

Explanation can serve two purposes: to help people to trust the system which generates an outcome through better understanding how it works, and to help people feel that the system is competent to carry out a function through providing credible justification for the outcome generated. In helping people to trust a system, it is important to show them how it works so that they do not feel that information is being kept from them. In some instances where consumers have queried a company about a decision made by AI which has affected their credit score, their ability to obtain a mortgage, or other life plans, companies have struggled to explain how the algorithm decides, given the vast number of data inputs and calculations which may take place in assigning a score.

Thus, one proposed aid to transparency is the introduction by developers of a "T switch" for transparency within a model. Corporate decision-makers could decide how opaque a particular algorithm could be and insert a "human in the loop" into the decision-making process. They could thus leave some routine decision-making to the algorithm acting autonomously. Still, they could require human input in a situation that has been identified as having significant real-world consequences. (For example, an AI would have the ability to offer you a pizza coupon while you are browsing online, but not the ability to independently show you ads for political candidates.) Other tasks which might require intervention might be the interpretation of law enforcement video footage, for example.

Combatting Disparate Impact Through Regulation

While developers thus have a responsibility to combat disparate impact in how they conceptualize, build, and apply their models, there are also legislative initiatives at local and federal levels to combat disparate impact through creating new regulations.

In 2019, Joy Buolamwini, founder of the Algorithmic Justice League, testified before the US House of Representatives Committee on Science, Space, and Technology. She argued that the proliferation of artificial intelligence in all areas meant that individuals often could not avoid being subject to algorithmic decision-making. Therefore, she said, governments must ensure that these decision-making algorithms are applied fairly. And since minority citizens may be more likely to be victimized by an imperfect algorithm, or biased judgment, it may be essential for members of such groups to have access to legal remedies, such as querying the algorithm.

In algorithmic justice, the European Union is in the lead in regulation. In February 2020, the District Court of The Hague, Netherlands, ruled that the System Risk indicator algorithm could no longer be used to flag individuals who were most likely to commit benefits fraud. This program, implemented in 2014 by the Dutch Ministry of

Social Affairs, used 12 categories of government data (including tax records, land registry files, vehicle registrations, etc.) collected from people living in low-income neighborhoods in several Dutch cities. The Dutch court determined that the ministry had violated the European Union's General Data Protection Regulation (GDPR) by collecting this data on this scale without informing the residents and further stated that applying an algorithm that "predetermined" that individuals with these characteristics were more likely to commit welfare fraud constituted a human rights violation.

In the European Union, regulations recognize the existence of so-called "indirect discrimination" – or a situation where a provision that appears to be neutral creates a particular disadvantage for a group that belongs to a protected category. In such a situation, the onus is on the people applying the provision to prove that it has a legitimate aim and that the means of achieving that aim are appropriate and necessary. This ruling may establish a precedent elsewhere in the European Union regarding the use of such risk prediction algorithms if it can be shown that they create indirect discrimination and that their use is not necessary or appropriate (Burack, 2020).

A Preemptive Approach to Preventing Bias and Disparate Impact

By asking developers to consider the possibility of bias being present in their models, as well as asking corporations to consider whether their use of a model will create disparate impact, the emphasis has shifted from addressing problems of bias and disparate impact once they are identified to instead preventing such issues from emerging in the first place. (That is, just as developers can design for security, they can also design for equity and transparency – from the beginning of the design process.)

A consensus seems to be emerging (at least in Europe) that the best way to combat disparate impact is thus not to attempt to fix or repair the results once they have already occurred but instead to have a formalized process (called a disparate impact audit) in which programs are scanned for disparate impact before they are launched. Currently, the European Union's General Data Protection Regulation requires data controllers (those with a responsibility for overseeing data activities within a corporation or government unit) to conduct a data protection impact assessment (DPIA) – under certain circumstances – to make sure that disparate impact is not occurring as the result of the application of an algorithm. The GDPR requires impact assessment if a practice is likely to result in an elevated risk to the rights and freedoms of natural persons, especially when using new technology or when organizations undertake fully automated decision-making, which is likely to significantly affect those who are subject to this decision-making. Furthermore, Article 22 of the GDPR states that "the data subject shall have the right not to be subject to a decision based solely on automatic processing called profiling" (Borgesius 2018). Thus, decisions like a bank denying credit or the government denying social benefits like pensions would all require a "human in the loop" rather than the mere application of an algorithm.

In the United States, the US Congress is still considering the Algorithmic Accountability Act, which requires companies to study their algorithms to identify bias and correct problems (Holstein et al., 2019).

Inequities, Injustice, and Dynamic Pricing

Transparency as a value has become more important to citizens who are the subjects of specific decisions (such as the decision to deny a bank loan). Citizens have also demanded greater transparency in understanding the algorithms that major data platforms use to offer goods and services to consumers through showing targeted ads, for example. Both Google and Facebook have been accused of engaging in discriminatory practices. Until recently, Facebook allowed companies to purchase targeted ads on their site to target people based on their sexual orientation and race. As a result, they were accused of aiding landlords who wished to engage in housing discrimination through, for example, not renting to Black or Hispanic families in the United States.

Box 8.2 Critical Issues: Policing and the Ethics of Body-Worn Cameras (BWC)

In March 2020, police in the US city of Minneapolis, Minnesota, attempted to arrest an African American man named George Floyd. In the ensuing arrest, Officer Derek Chauvin was observed kneeling on the neck of George Floyd for 8 minutes and 46 seconds as he attempted to restrain him. Across the United States and the world, observers viewed footage obtained from a body-worn camera and footage compiled by bystanders to the event. Floyd died at the scene, and Chauvin was later charged with murder (Blewer, 2021).

For many in the United States, the event showed how BWC footage could be used to create police accountability and transparency. The availability of video footage allowed those in court to judge whether the force applied was appropriate and whether proper procedures had been followed. For citizens of Minneapolis, the event allowed citizens to call for change in the police force that many viewed as insufficiently responsive to the community's needs, including the African American community.

In November 2019, a similar incident occurred in Australia's Northern Territory. Nineteen-year-old Umanjayi Walker was shot dead by a police constable. His death was recorded on camera, and the trial was still ongoing as of this writing (Blewer, 2021).

But what are the goals of using BWCs in community policing, and are these goals being met?

What Are BWCs For?

The use of BWCs is widespread, but a consensus still does not exist about the goals of introducing BWCs in police work. Thus, debates exist about whether BWCs are adequate and whether technology is sufficient to solve problems like inequities in policing, police brutality, and how the public views the police.

BWCs can have three different types of effects when they are used.

First, mandating that officers utilize BWCs can affect the behavior of the officers and first responders and those who are apprehended. As Blewer (2021) notes, "BWCs, worn by officers, make visual recordings of what the wearer does, sees and hears in the execution of their duties" (p. 1181). However, the evidence regarding whether such recordings can change behavior is mixed. Some studies suggest that officers are less likely to use force and that situations are less likely to escalate when police wear BWCs.

Secondly, the use of BWCs can increase transparency in a community. People are more likely to have faith in and trust their police departments when they are provided with complete information (including the ability to view videos of individuals engaged in policing) about what transpires in their communities, especially in a conflict.

Next, BWCs can provide helpful additional information, including evidence, in court proceedings. Disputes about drug possession, whether someone resisted arrest, or whether or not someone was read their rights can easily be alleviated by referring to video evidence obtained.

What Are the Downsides to the Use of BWC?

The Hawthorn Effect is a well-known psychological principle that states that people behave differently when they know that they are being watched, or that laboratory subjects behave differently when they are the subjects of an experiment. People might be more cooperative with an interviewer in an experiment than they usually are, or they might modify their behavior in response to social pressure. Some advocates believe that the Hawthorn Effect can be used to induce police personnel, for example, to behave more cooperatively and less aggressively towards the people they are apprehending. They believe that mandating that officers wear BWCs can deter officers from engaging in aggressive behavior – since they know they are being watched (Clare et al., 2021).

However, some police union advocates are concerned that the ultimate result might be one of "de-policing," in which law enforcement officials restrain their behavior to the point that they are ineffective, fearing the repercussions if they were later to be called to account for a decision to use force or behave aggressively.

In addition, the regulations regarding ownership of BWC footage and the rules regarding storing, sharing, and copying this footage are still in a state of flux. In the United States and Australia, individual police districts are viewed as owning this footage. This means that another entity, like the US Department of Justice (who often investigates cases that might be viewed as tainted by racial or other forms of bias), would have to request access to such footage legally and might have to negotiate with the police district regarding what is made available (Blewer, 2021).

Applying the Lenses

From an ethics standpoint, one might make a practical argument, suggesting that if the overall effect of mandating that officers wear BWCs is an increase in transparency and public trust, along with the deterrent effect upon police officers who might otherwise use violence, then the overall good produced by mandating BWCs might outweigh any other concerns. A deontological lens would suggest taking the viewpoint of the least influential person in the scenario. If mandating BWCs can protect the least powerful subjects from potential police aggression, this lens would again support their use.

However, the virtue ethics lens suggests a slightly different outcome. Here, one could argue that ultimately a community should wish to hire law enforcement officers who are well-trained, honorable, and respectful of community members. In this instance, merely mandating that officers wear a sort of "electronic leash" to restrain those members who might have inherent violent tendencies (or racist attitudes) is not a good way of creating a moral police force. Instead, a community might wish to revisit how personnel are recruited, hired, screened, and trained, to filter out those who are perhaps pursuing a career in law enforcement for the wrong reasons.

In addition, some analysts suggest that video evidence is not always clear-cut. There may still be grey areas where the video is subject to interpretation. For example, a factor like the specific camera angle might ultimately distort what viewers see. They urge caution in blindly assuming that "the camera doesn't lie."

Competing Values

As noted, the use of BWCs can potentially increase the accountability which police officers have for their policing decisions and their interactions with the community. In addition, the use of BWCs can increase transparency and, hopefully, trust and confidence in the community in the police force.

Another value to consider is that of equity. Here, BWCs can help produce more equitable and fair outcomes in court by providing more objective evidence of what transpired during an arrest. In addition, the use of BWCs can produce equity by assuring that law enforcement officers treat all suspects the same, regardless of factors like gender and race. In addition, some police districts have argued that using BWC footage can help them monitor officers' behavior, leading to more equitable personnel policies and decisions in areas like layoffs, firings, and promotions (Clare et al., 2021).

However, as noted earlier in this text (see Box 5.2, "Critical Issues: Ethics of Worker Surveillance"), creating equity, accountability, and transparency can sometimes create situations where workers (like law enforcement officers) have fewer privacy rights. Strict regulations must be put into place governing who should have access to police surveillance videos, under what conditions, and how such materials should be shared, stored, and copied.

Bibliography

Blewer, R. A. (2021). Every move you make . . . every word you say: Regulating police body-worn cameras. *University of New South Wales Journal*, *44*(3). Retrieved June 1, 2022, from www.unswlawjournal.unsw.edu.au/article/every-move-you-make-every-word-you-say-regulating-police-body-worn-cameras/.

Clare, J., Henstock, D., McComb, C., Newland, R., & Barnes, G. (2021). The results of a randomized controlled trial of policy body-worn video in Australia. *Journal of Experimental Criminology*, *17*(1), 43–54.

A Google program that selected advertisements for job seekers was accused of showing ads for low-paying jobs to female job seekers and ads for higher-paying jobs to male job seekers. In addition, some companies are accused of practicing age discrimination by "screening out" older job seekers by not showing them job advertisements.

Bellanova et al. (2021) use the term *datafied social relations* to refer to how our connections are "curated," often without our knowledge by algorithms that decide which media to show us, which connections to other individuals to suggest, and other information.

Here, one can identify two ethical issues: First, individuals were not treated equitably since they had different online experiences with different opportunities suggested to them based on gender or race. Some individuals were the subject of discrimination, while others were not. Secondly, people were often unaware that they were the subjects of discrimination and inequity. Platforms have been criticized thus for helping employers and landlords who wished to engage in discriminatory conduct and for engaging in opaque practices that people could not speak back to, since they were unaware they were happening.

Some platforms have also been accused of offering the same product to different consumers at different prices. For example, a frequent shopper on a site might be offered a better price. In one case in the United States, clients for an SAT tutoring company, The Princeton Review, were charged different prices for the same course, depending on what region of the country they lived in.

Ethicists currently disagree about the moral legitimacy of such practices. Here we can see how competing values come into play. Someone whose primary value orientation is efficiency might argue that as long as everyone can purchase a good they need, it should not matter if individuals pay different prices. Others, however, argue that it is essential that everyone have the same chance to purchase a good or service at the same price to have an equitable online experience.

In her work, Antoinette Carroll, a tech professional, writes that "systems of oppression, inequality, and inequity are by design . . . therefore, they can and must be redesigned." Her equity-centered community design framework asks designers to consider factors like history, context, power dynamics in the needs of those who will live with the outcome of technology (Carroll, 2020).

This debate is still emerging, and regulation is again lagging behind the new and novel ethical issues apparent here.

How Technology Can Create and Shape Unjust Patterns

A final ethical issue to consider in this chapter is whether we, as developers, computer scientists, and citizens, merely use technologies like Big Data – or whether a feedback loop exists in which technology shapes us. Today, collecting, organizing, and analyzing data follow specific patterns. We ask some questions, and not others, because the data lends itself to asking some types of questions more readily. For example, Big Data seems to be particularly good at predicting and identifying risk factors. Therefore, a successful research project for cybersecurity analysts might focus on examining data about previous cyberhacking attacks to identify patterns regarding the sorts of attacks that are likely to occur and the vulnerabilities a system might have.

But some researchers today argue that because we can ask and answer questions about risk so quickly, we might be tempted as researchers to overemphasize the study and prediction of risk factors over asking other types of questions. As a result, we risk creating a world in which we often think about individuals or groups regarding the criminal risk they pose or where we frequently sort people and organizations into groups based on our perception of their potential risk. Green and Chen (2021) argue that today, AI and Big

Data have allowed us to apply algorithms to determine risk in many areas of life – from predicting which neighborhoods in a city pose the most significant criminal threat, to asking which types of parents are most likely to abuse their children, to considering who is most likely to develop a disease. They also point out that because quantitative data and the visualizations we create with it are so compelling, we might be tempted to overrely on this data, sometimes overriding what we know personally or professionally about an issue. Here, they describe how judges, in some instances, will override their knowledge and instincts about a criminal case and instead defer to sentencing software which generates a prediction about the risk that someone facing incarceration will go out and re-offend if they are released into the community.

Furthermore, it has been suggested that our reliance as employers, researchers, developers, and users on Big Data may lead us to adopt an ethical position in which we regard individuals themselves as mere "inputs." It is easy not to see a worker, for example, as a human who is owed certain rights and dignity – and instead, to merely view them as either productive or unproductive, quick to learn or slow to learn, or as someone who presents a cost to their employer. (In this view, workers might be mere "assets" to be deployed to create a product.) Such an approach fails to see people as individuals and may not fully account for their contributions to an organization that is not easily quantified (Royakkers & Olsthoorn, 2014). Thus, a human resources employee who is asked to identify the bottom 10 percent of employees in terms of their productivity might turn to a data analytics program that quantifies such measures as the number of customer service calls completed during a shift – without considering each employee's unique skills and gifts.

Corporations React to Problems of Disparate Impact

In recent years, corporations have begun to speak out about ethical issues related to surveillance and disparate impact. In the wake of the global 2020 Black Lives Matter protests, IBM ended the development of facial recognition technology, while Microsoft committed to not selling this technology to law enforcement. And Amazon placed a one-year moratorium on police use of this technology.

However, it is unclear to what degree a platform should be held liable for the practices users engage on that platform. Just as a firearms manufacturer is not usually held liable if someone issues a firearm to engage in murder, some ethicists feel that Amazon, for example, should not be held liable if another company downloads its data and uses it in an unethical way. While the increased use of Big Data and algorithms can open the door to an increase in ethical issues like disparate impact, it is often not clear whose specific responsibility it is to address these unethical practices, nor is it clear what specific responsibilities each member of a data partnership might have in such a situation.

Competing Values

Today, developers in artificial intelligence have identified several ethical values that may come into play when addressing problems like bias and equity. The term zero-sum refers to a problem with only one solution set. To achieve one outcome, the player has to abandon any chance of receiving any other outcomes or even commit to receiving a negative outcome on one value to receive a positive outcome in another. In thinking about the ethics of developing machine learning algorithms, developers may find themselves thinking about trade-offs or opportunity costs.

Here, some developers argue that the most crucial goal a developer should have is to create an algorithm that is as accurate as possible. An algorithm should correctly identify a likely security risk or who is likely to abuse their children with as high an accuracy score as possible. Thus, they ask, should one worry about other values, like transparency or equity, if achieving these outcomes might mean that the model is less accurate, runs significantly slower as a result, or is more expensive to run?

Developers have also argued that the best way to get an accurate model is to run and apply it as much as possible to feed the model more data and optimize its running. Thus, they argue that it may be necessary to run a model before it has been thoroughly vetted about issues like equity. At the same time, they argue, it is better to feed a model as much data as possible, even if the data is historic and reflects existing societal inequities (such as identifying men as doctors and women as nurses). Here, they argue that this may be the only data available, and more data is better than fewer data.

Currently, it can seem like our machine learning models are voracious learners hungry for new data. As a result, developers have sometimes made ethical trade-offs such as training a model on data whose provenance may be questionable, or where it is unclear who owns the data and thus can grant permission to give it. (For example, people's abandoned family photos on a defunct service like Flickr might be used as training data for a machine learning model.)

Here, an ethicist who is also pragmatic or oriented towards real-world issues might suggest that not all ethical errors in machine learning are equally severe. Instead, they might argue that in considering the problem of AI decision-making bias and how to address it, the developer should consider the impact of the model's outcomes. (Is the AI deciding to show you an advertising video, or is it deciding to send you to jail?) They should also consider the types of decisions being made by an AI. Here people may feel okay about having an AI be just an image to make it more readable, but not about having an AI read your medical scans and dispense medical advice.

A utilitarian would note that customizing an algorithm or adjusting it to be most accurate for different groups can add layers to an algorithm, pushing up development costs. Customization can also raise production and monitoring costs, increasing organizational complexity and overhead. The most equitable algorithm might not be the most accurate, the cheapest, or the most efficient.

In addition, an emphasis on equity can sometimes lead to a trade-off regarding who can access a program and under what conditions. Developers may choose not to develop or sell to a market if the regulations are too complex or expensive to implement. Companies wishing to do business with Europe might decide against doing so if the requirements imposed by the GDPR are seen as too demanding – encompassing system audits, AI monitoring and diversity awareness training, and the development of materials aimed at increasing transparency regarding algorithms. Furthermore, companies may oppose transparency out of a fear that proprietary information or trade secrets will need to be revealed. And nations which do not have transparency requirements, like China, may be able to "out-compete" other nations that are bound by stricter protocols in this regard.

Thus, in considering equity, access, and transparency problems, one can encounter issues with competing values – accuracy and efficiency.

CHAPTER SUMMARY

- Those who design new technologies are wise to consider diversity, equity, and inclusion principles at all stages of the design process.
- Due to issues of data bias or incomplete data, algorithms may end up having a disparate impact on some social groups – including women and ethnic, racial, and sexual minorities.
- One leading way to create equity and justice in situations where algorithms are serving as decision-makers is through mandating practices of transparency. In Europe in particular, individuals have the right to understand how a decision was made or an algorithm was applied.
- Many commercial providers today provide suites of tools which can be used to test to see if datasets contain bias or if an algorithm is likely to return a biased result.
- Those who design new technology may find themselves having to consider competing values – like efficiency and accuracy, as well as equity and inclusion.

DISCUSSION QUESTIONS

1 **Writing an Algorithm to Distribute Scarce Resources**

Read the following article: Lynn A. Jansen and Steven Wall. "Weighted Lotteries and the Allocation of Scarce Medications for COVID-19," *The Hastings Center Report*, January-February 2021. Vol. 51, Issue 1: 39–46. The authors describe how a weighted lottery system could allocate scarce medical resources more efficiently to those most likely to benefit from them.

You are a computer engineer who has been asked to create a program that would classify individuals into priority groups for receiving medications based on the criteria listed in the article.

- What ethical problems do you anticipate?
- Do you regard this algorithm as fair?
- What, if anything, could be done to make the criteria transparent to those who will be subject to its decisions?
- Do you think a computer alone should decide who belongs to which priority group? Or should there be a "human in the loop"? Why or why not?
- Are you worried about the disparate impact on any group?
- If you belong to a cultural or religious group with its own moral principles (i.e., respect for the elderly, sacredness of all human life, etc.), would a weighted lottery work, or does this conflict with these principles in any way? Why or why not?

2 **Using Facial Recognition for National Security**

In the report "The Centaur's Dilemma: National Security Law for the Coming Artificial Intelligence Revolution," written for the Brookings Institution think tank in Washington, D.C. (Baker, 2020), Baker writes that "According to the Government Accountability Office (GAO), Since 2011, the FBI has logged more than 390,000 facial recognition searches of federal and local databases, including state DMV databases with access to 641 million face photos."

The FBI has said that its system is 86 percent accurate.

- What accuracy percentage would or should be necessary before deploying such a system?
- Apply the paradigms here. How might a practical, a deontological, and a virtue ethics stance help to answer this question.
- What values would you use in making your decision? Consider the value of benevolence, nonmaleficence, equity, and justice.
- Extra credit: Is this 86 percent accuracy rate likely to be randomly or evenly distributed in the population? Does it matter if some people face a higher risk of inaccurately identifying?

3 **Thinking About Competing Values**
Look at the graphic in Figure 8.1. Of the five values listed – access, efficiency, equity, transparency, and fairness – explain what values you think may be opposed to one another and which might be complementary when thinking about the applications of algorithms. Can an algorithm be *both* equitable and transparent, both efficient and fair? Think of specific examples to justify your assertions.

4 **Thinking About User-Centered Design**
Let's consider the idea of user-centered design.

- Is the social media platform like Twitter as currently designed, *created* in such a way that users are empowered more than corporations or bad actors?
- What could be done to *design* the space in such a way that it was open and transparent, rather than opaque, so that it served as a space for community discussion rather than polarization, so that people were more easily able to find high-quality, reputable sources of information, etc.?

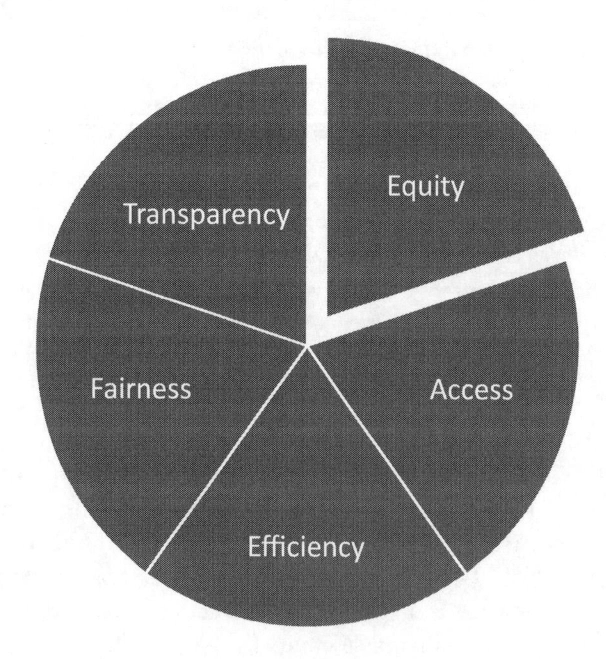

Figure 8.1 Competing Values

- Do you think corporations would be amenable to adopting practices related to "designing for democracy"? Why or why not? What might such practices look like?
- Is designing for democracy a reasonable goal, akin to adopting practices like designing for privacy, security, and accessibility?

RECOMMENDED RESOURCES

Documentary, *Coded Bias* by filmmaker Shalini Kantayya. Deals with racial and sexist bias in face-recognition software and other AI systems (available on Netflix).

Kirkland, K. "Wearable Technology Helps You Navigate by Touch." Retrieved from www.youtube.com/watch?v=xnQB9Y77PXE.

McIlwain, C. D. (2019). Black software: The internet and racial justice, from the front to black lives matter. Oxford: Oxford University Press.

NYU Center for Critical Race and Digital Studies: https://criticalraceanddigitalstudies.com

Students may wish to visit this website: http://aif360.mybluemix.net (accessed June 21, 2022) to learn about AI Fairness 360, a set of tools to be used with the IBM Watson Machine Learning tool. These tools can examine one's datasets and models for bias by running several different tests. The site includes a web demo and several videos.

References

Baker, J. E. (2020). *The centaur's dilemma: National security law for the coming artificial intelligence revolution*. Washington, DC: Brookings Institute.

Balayn, A., & Guises, S. (n.d.). *Beyond debiasing: Regulating artificial intelligence and its inequalities*. Brussels: European Digital Rights.

Bellanova, R., Irion, K., Jacobsen, K., Ragazzi, F., Saugmann, R., & Suchman, L. (2021). Toward a critique of algorithmic violence. *International Political Sociology*, *15*(1), 121–150.

Bichell, R. E., & Anthony, C. (2021, June 6). For Black kidney patients, an algorithm may help perpetuate harmful racial disparities. *Washington Post*. Retrieved from www.lipstickalley.com/threads/for-black-kidney-patients-an-algorithm-may-help-perpetuate-harmful-racial-disparities.4567175/#:~:text=An%20algorithm%20doctors%20use%20may%20help%20perpetuate%20such,patient%E2%80%99s%20age%20and%20sex%2.

Biddle, S., & Saleh, M. (2021, August 25). Little-known federal software can trigger revocation of citizenship. *The Intercept*. Retrieved June 4, 2022, from https://ground.news/article/little-known-federal-software-can-trigger-revocation-of-citizenship.

Borgesius, F. Z. (2018). *Discrimination artificial intelligence and algorithmic decision-making*. Strasbourg: Council of Europe.

Burack, J. (2020, November 24). Addressing algorithmic discrimination in the European Union. *A Path for Europe*. Retrieved from https://pathforeurope.edu/addressing-algorithmic-discrimination-in-the-european-union/.

Carroll, A. (2020, October 21). Social justice and equity by design. *3 BL Media*. Retrieved June 3, 2022, from www.3blmedia.com/news/social-justice-and-equity-design.

Criado-Perez, C. (2020). *Invisible women: Data bias in a world designed for men*. London: Penguin Books.

Dankwa-Mullan, I. (2020, June 24). Eliminating bias in health data science. *Think IBM Blog*. Retrieved June 3, 2022, from www.ibm.com/blogs/watson-health/eliminating-bias-in-health-data/.

Debrusk, C. (2018, March 26). The risk of machine-learning bias (and how to prevent it). *MIT Sloan Blogs*. Retrieved June 3, 2022, from https://sloanreview.mit.edu/article/the-risk-of-machine-learning-bias-and-how-to-prevent-it/.

Gilliard, C., & Golumbia, D. (2021, July 6). Luxury surveillance. *Real Life Magazine*. Retrieved from https://reallifemag.com/luxury-surveillance/.

Holstein, K., Vaughan, J., Daume, H., Wallach, H., & Miroslav, D. (2019, May 4). Improving fairness in machine learning systems: What do industry practitioners need? *Computer-human interactions 2019 paper*. Retrieved from http://users.umiacs.umd.edu/~hal/docs/daume19fairness.pdf.

Konopasky, A. (2021). Pre-employment tests of "fit" under the Americans with disabilities act. *Southern California Review of Law and Social Justice*, *30*(2). Retrieved May 31, 2022, from https://gould.usc.edu/students/journals/rlsj/issues/assets/docs/volume30/spring2021/Konopasky.pdf.

MacNish, K., Fernandezinguanzo, A., & Kirichenko, A. (n.d.). Smart information systems in cybersecurity: An ethical analysis. *Orbit Journal*, *2*(2). https://doi.org/10/2927/orbit.v2i2.105.

Nelson, C. A. (2017, November 20). Moving away from data invisibility at tribal colleges and universities. *Higheredtoday.org*. Retrieved June 3, 2022, from www.higheredtoday.org/2017/11/20/moving-away-data-invisibility-tribal-colleges-universities/.

Proctorio's awful reviews disappear down the memory hole. (2021, September 4). *Pluralistic*. Retrieved June 3, 2022, from https://doctorow.medium.com/proctorios-awful-reviews-disappear-down-the-memory-hole-5eb3e11bdbaf.

Reynolds, M. (2017, April 1). Bias test to keep algorithms ethical. *New Scientist*, p. 10.

Royakkers, L. M. M., & Olsthoorn, P. (2014) Military robots and the question of responsibility. *International Journal of Technoethics*, *5*(1), 1–14.

TechTarget Contributor. (n.d.). Algorithmic transparency. *Techtarget.com*. Retrieved June 3, 2022, from www.techtarget.com/searchenterpriseai/definition/algorithmic-transparency.

Vaughan, A. (2021, March 20). Passport tool fails with dark skin. *New Scientist*, p. 19.

Wing, N. (2014, June 25). Seven ways that you (yes, you) could end up on a terrorist watch list. *Huffington Post*. Retrieved June 1, 2022, from "Seven Ways that you (yes, you) could End up on a Terrorist Watch List." Huffington Post, June 2.

Wong, Q. (2019, March 27). Why facial recognition's racial bias problem is so hard to crack. *cnet.com*. Retrieved June 1, 2022, from www.cnet.com/science/why-facial-recognitions-racial-bias-problem-is-so-hard-to-crack/.

9 Big Data and the Ethics of Cybersecurity

LEARNING OBJECTIVES

At the end of this chapter, students will be able to:

1. Describe the major ethical issues associated with data storage and data sharing when working with Big Data – including issues of informed consent
2. Describe the major ethical issues associated with using Big Data for decision-making, including issues of diversity, equity, and inclusion
3. Define key terms associated with Big Data, including qualitative and quantitative data, structured and unstructured data, raw data and open-source data, customized and aggregate data
4. Compare and contrast ethical positions regarding whether data is best understood as an individual or a collective good

Introduction

Is there such a thing as the ethics of Big Data? In thinking about Big Data, we can identify several areas of ethical concern. First, we can consider the ethical choices that practitioners – including data engineers, data analysts, and cybersecurity professionals – face in working with Big Data. Next, we can examine how Big Data is collected, stored, analyzed, and used to make decisions. Here we can consider how Big Data can tell us about ourselves as humans and even how it can interact, nudging us to behave in specific ways. Next, we can build upon the themes introduced in Chapter 8 on diversity, equity, and inclusion to examine the relationship between society and Big Data – considering how data does and does not represent those in society, how it can empower and disempower those in society, and how it can change how we perceive situations and take actions as a result.

Throughout this chapter, we consider the notions of rights and responsibilities, including the responsibilities of those who work with Big Data (either individually as practitioners, or as corporation leaders making policies), as well as the relationships and obligations of individual data users and their relationships to corporations that might buy or sell their data.

What Is Data?

Scientists have always collected data – from making observations in a laboratory to compiling the survey results administered in person, via the telephone, or online. But what

DOI:10.4324/9781003248828-12

exactly is data? One definition tells us that "data are facts, observations, or experiences on which an argument or theory is constructed or tested" (Cox & Verbaan, 2018, p. 23).

Data comes in several different forms, as we will see in this chapter. *Quantitative data* refers to information that can be manipulated using mathematical techniques called quantitative analysis. Quantitative analysis can include statistics and other techniques for working with large datasets, such as SQL queries. Data analysts use programs like Python and R to work with quantitative data. Quantitative data might include information like the percentages of people who purchase a particular product and include elements like answers to a multiple-choice survey question.

In contrast, *qualitative data* refers to data that is not easily subject to quantification, including measurement and the application of mathematical and statistical tests. It may include answers to open-ended survey questions and interviews and some types of information compiled from historical records, court transcripts, and other documents. Scientists use different methods for working with qualitative data.

In addition, today, data scientists often distinguish between *structured data* and *unstructured data* rather than speaking of qualitative and quantitative data. Structured data, which is quantitative, can easily be labeled and put into a structure like a spreadsheet. This labeling and uploading of data (or extracting, transforming, and loading data) can now be performed by AI agents trained using machine learning.

Unstructured data, in contrast, can be more challenging to work with. Unstructured data is the data that users often produce by making and uploading a video to a site like YouTube or TikTok. This data is often difficult to categorize and work with, and new techniques (like making a word cloud to judge user sentiment) are being developed.

Here, we can also distinguish between the two types of data – structured and unstructured – by noting that structured data is "made data" obtained through experiments and surveys, while unstructured data is often produced organically – generated, not for research purposes. Such data is also described as "found data." This distinction becomes important when we begin to consider both data and research ethics – since those who take a survey to produce "made data" must consent to participate in this data-gathering effort. In contrast, when researchers begin to use "found data," issues of consent arise since the data was not initially produced for research purposes. Found data may also be more error-prone. Finally, researchers who collect "made data" adhere to strict protocols regarding protecting the confidentiality and anonymity of research study participants. At the same time, found data may be more readily identifiable in some instances. For these reasons, using unstructured and found data for research purposes may be ethically challenging (Garfinkel, 2018).

Data thus refers to numbers, descriptions, lists, music, and other aural elements produced, and visual elements like photos and drawings. *Raw data* is collected before transformation and conversion operations, and data cleaning. *Open-source data* is collected and then shared with other researchers for free. Some government programs require that data collected by the government itself or researchers using a government grant be made open source. Other types of data, such as that collected by a commercial company, may be labeled as proprietary or private, and companies may resist sharing this data since they do not wish to make trade secrets public inadvertently.

What Is Big Data?

Big Data is a relatively new term. It was first used by researchers from the National Aeronautics and Space Administration in 1997. Big Data refers to "massive volumes of information collected through technological means, accumulating at such a velocity that is continually innovating information processing is required to keep pace." Today, the amount of information available to information consumers doubles every 18 months. Someone sitting at home at their computer has access to more information than people in the Middle Ages could even dream of. It is anticipated that human knowledge will double every 12 hours within a few years (Jurkiewicz, 2018, p. 547).

While researchers in the past struggled with collecting enough data to carry out a research project, today, researchers are often inundated with data. They face challenges in creating technologies that can deal with the massive volume of data and sorting through the data deluge to find the most accurate and relevant data for their needs. As the internet has grown and changed, making it possible for researchers to communicate better and compile information, many pundits have made optimistic, even giddy pronouncements regarding how science would now be able to solve all the world's problems – from solving the energy crisis, to allowing humans to travel to Mars, to increasing the likelihood that we will find a cure for diseases like cancer.

Qualities of Big Data

What distinguishes Big Data, then, from other types of data? Researchers today use the "Five V's" to describe what makes Big Data unique. First is the volume of data contained in Big Data. As noted earlier, more data is produced today than in previous centuries. Big Data is also unique for the variety of information that it contains. In particular, creating unstructured data like transcripts of YouTube videos, the uploading of photographs, and the ongoing collection of data from Internet of Things devices means more types of data are collected, archived, and analyzed today than ever before. Next, Big Data differs from other data types due to the velocity or speed it produces, shares, and analyzes. While the lead time for the publication of an academic article might be two years from when an academic submits it to an editor to when it sees publication, today, data can be collected, analyzed, and disseminated quickly, enabling the carrying out of activities like the almost real-time tracking of a pandemic.

Next, Big Data is characterized by its integrity or truthfulness. Because there is so much data produced today by so many different sources, issues may arise regarding the truthfulness of knowledge disseminated and claims made based on data. Today, data analysts and data engineers are involved in developing measures to find and identify the source of data collected and verify its integrity.

Finally, Big Data is valuable. Information collected by social media platforms like Facebook, which contains records of what individuals like to view on television and purchase in stores, and what events are most important to them, can be repackaged and resold to advertisers and merchants. As Jurkiewicz notes, data can travel across contexts. Data about the weather in local areas collected by volunteers can then be repackaged as a commercial source for weather forecasters. They can even be influential in decisions about future agriculture on the stock market (Jurkiewicz, 2018, pp. 27–29).

Ethical Issues Related to Big Data

Because Big Data is a big business, ethical issues related to data ownership are particularly salient. Today, we are just beginning to ask questions like, who owns my data, or the data I generate through my online activity? Who can decide how that data is used, resold, stored, and aggregated with other data?

Here, the terms *data exhaust* and *data stream* are essential to understand. Data "exhaust" is a term for the trail you might produce while browsing online, as your browsing behavior can offer valuable insights to someone wishing to sell you a product. You are possibly not even aware that you are producing data exhaust and may not think much about the data you are generating or how it might be used later and by whom.

Box 9.1 Critical Issues: What Is Wearable Technology?

Wearable technology is defined as technology that is designed to be worn, often for the long term, by an individual user for the purposes of monitoring behavior and collecting data. A wearable device might also interact with a user by providing feedback about the user's behavior. Currently, the most well-known wearable is a fitness tracking device, like an Apple Watch. Approximately 20 percent of the US population uses such a device.

But what are the pros and cons of using such devices, and what sorts of ethical and legal issues do they raise?

Benefits of Using Wearable Devices

Wearables can be used to improve individual health behaviors, to produce better data about human physiological performance, and even to monitor compliance of individuals and groups for the purposes of public health.

On an individual level, wearables can engage in *hyper nudging*, through providing feedback to users in order to move them towards a desired behavior (such as moving up to 10,000 steps a day). Graphic interfaces might appear on a device which encourage a user to engage in socially beneficial behaviors such as getting enough sleep, exercising daily, or cutting back on social media usage (Lanzing, 2019). A feature that listens to your tone and voice and prompts you whenever your voice becomes loud or hostile could even help you to improve your relationship with your children or your spouse by gradually, over time, teaching you to yell less or use other methods of conflict resolution.

In group settings, the use of wearables may help a college athletics team to improve their performance or may even serve as a valuable source of athletic performance data which researchers can use to suggest ways that we can all improve our fitness. Military researchers might also collect data in an attempt to improve military performance in combat situations.

Public health researchers have also developed wearables in which individuals might be asked to swallow a pill which contains a form of digestible sensor. Someone who is wearing a tracking device might therefore be able to send data indicating that they are complying with a treatment regime through taking such a pill daily.

Finally, new wearable devices are being developed for users with disabilities, such as systems which allow those who are visually impaired to use *haptic systems* to navigate the world by touch (Kirkland, 2019).

The use and deployment of wearable technology therefore supports such values as access and equity, responsibility, health, and personal improvement. In addition, using a utilitarian lens, one might argue that the goods produced by offering or even mandating the use of wearable technology – in the form of increased gathering of valuable medical information, increased compliance with public health rules and procedures, and increased individual health – far outweigh any risks posed by either voluntary or compelled adoption of this technology.

What Are the Downsides?

However, not everyone would agree with this utilitarian approach. In thinking about wearables, we can also identify several associated ethical and legal issues. On an individual level, it has been suggested that asking or encouraging people to use wearables can erode their personal autonomy. Here, Lanzing (2019) warns that hyper nudging self-tracking technology does, however, have the ability to compromise the user's autonomy as well as violate information and decisional privacy. She argues that while you may believe that you are choosing autonomously to put in enough steps to fill up the circle on the Apple Watch, in reality the graphical interface is doing some work in steering you towards a desired behavior, even without your knowledge and consent. Furthermore, she notes that such devices often appeal to our competitive instincts by presenting our individual data to us within the context of a similar demographic group (i.e., telling someone that they run faster than 90 percent of other women of their age and fitness level). While doing so may incentivize some users to improve their performance, she notes that in order to present such data to users, other users' data is stored and compiled, often within their specific knowledge and consent.

She introduces the notion of *decisional privacy*, defined as the right against unwanted access such as unwanted interference in our decisions and actions. She argues that we each have a right to decisional privacy, and that designers therefore should aim to protect us from intrusion and interference in our decision-making and to provide us with the freedom to exercise autonomous or personal decision-making.

The Issue of Consent

Thus far, we have discussed wearables as a technology which users would adopt voluntarily out of an individual desire to improve their own performance. However, we can also conceptualize of many scenarios in which an outside actor – such as an athletic coach, an employer, or a law enforcement agency – might wish to incentivize you to improve your behavior. The technology therefore raises issues of consent. Could your employer demand that you wear a tracking device, so that the employer can see whether you are exercising daily, for example? And could public health officials demand that all citizens were behavior trackers?

Here deStefano (2017) argues that in a public health scenario, for example, those who are most likely to be monitored are likely to be the most vulnerable in our societies. While current scenarios include the possibility of inducing a population for the homeless community to wear medication trackers in order to monitor and lessen their risk of passing on communicable diseases like tuberculosis to the community, other scenarios might include monitoring whether people are taking birth control pills, or hormonal therapies sometimes prescribed to sexual offenders.

Women, people of color, and those who are in poverty are most likely to be tracked in such a way, he argues, while those who are wealthy and belong to mainstream cultural groups are not (deStefano, 2017).

The US Equal Employment Opportunities Commission, which monitors workplace diversity in the United States, has issued guidance regarding situations where workers might feel pressured to wear a tracking device in order to, for example, receive a more favorable rate on employer-provided health insurance. The EEOC guidance states that an incentive (like a reduction in health care costs) may be offered by an employee in exchange for wearing a device, but that an employer should not make employees feel pressure to disclose private medical information or disability status information.

In her work, McCarthy (2021) describes how wearables are increasingly commonly used in Division I US College athletics. She points to ethical issues associated with access and equity in describing how wearable technology can be used by college athletes to enhance their performance. Wearable devices can collect physical and physiological data about sleep, stresses, rest, and recovery. Sophisticated devices today can include heart rate monitors and GPS trackers.

Who Owns This Data?

While she worries that top-tier athletic programs may have an unfair advantage over less well-resourced programs that do not have access to such technology, she also raises programs that well-resourced athletic departments may face. She points out that many top Division I schools have struck deals with Nike athletics whereby wearable technology is provided in exchange for a student-athlete's data, raising issues of privacy and consent. Currently the National Collegiate Athletic Association (NCAA) allows players to wear search devices during games. However, the NCAA has not issued any specific guidance regarding who owns the data which is collected – whether it belongs to students' athletic departments or to an organization like Nike.

Ethical Responsibilities for Wearables Developers

How might those who wish to develop such apps proceed in ways that are ethical and in compliance with the ideas of their profession? Developers might implement principles related to privacy by design, and they might also cultivate an awareness by users of how their data is being used, stored, and shared. They might require users to consent to all data uses and also provide an orientation which helps users to understand how decisions like colors, fonts, and graphics all serve to "steer" them towards certain behaviors.

Bibliography

deStefano, M. (2017). Wearable biometric technologies and public health. *American Journal of Bioethics*, *17*(1), 79–81.

Kirkland, K. (2019, April 26). Video presentation: Wearable technology that helps you navigate by touch. *YouTube*. Retrieved from https://www.youtube.com/watch?v=xnQB9Y77PXE&t=9s.

Lanzing, M. (2019). Strongly recommended: Revisiting decisional privacy to judge hypernudging in self-tracking technologies. *Philosophy of Technology, 32*(1), 549–568.

McCarthy, C. (2021). Consider ethics of wearable technology and NIL. *College Athletics and the Law, 18*(8), 5–6. Retrieved September 26, 2022, from https://doi.org/10.1002/catl.30943.

A company might wish to know how long you spent on a site or which links you clicked on. Data streams are produced and may be utilized "downstream" by other clients who might also wish to glean insights from your online behavior. The question becomes, at what point as your data flows downstream does it cease to be your data, subject to the specific consent which you may have given regarding its use when you initially landed on a page or performed a search? and "what may or may not be done ethically with the data which you are producing?" (McCormack, 2016). That is, what are your downstream rights? At what point do you *stop* owning your data? Once it has been transformed? Do you still own it once it has been aggregated with multiple other datasets, based on the assumption that it is now no longer readily identifiable as you?

This chapter will consider the ethical issues associated with each phase of data collection – from the site visitor's original statement of consent to the point at which data is processed, where data is shared or sold, and when it is stored. Different types of individuals – from data architects to data analysts to marketing experts – may encounter this data at each stage, and the ethical difficulties may differ. Researchers must think about research ethics, respect for the people who supply the data, and the limitations they encounter based on what they identify as the goal of their research. Cybersecurity experts must think about the security principle or the requirement that data be kept safe.

Data Ethics and Cybersecurity Ethics

In this chapter, we distinguish between cybersecurity ethics and data ethics. Data ethics is a branch of ethics concerned with data and information – including how it is produced, configured, and stored. Data ethics researchers consider the ethical uses of tools like algorithms and professional codes.

Many of the ethical questions that cybersecurity practitioners ask overlap with those that data ethicists ask. Both groups ask questions like: "Who is my client?" and "What are the accepted norms (vs. regulations and rules) in a field that is rapidly changing?" Cybersecurity ethics queries may also overlap with research ethics more generally. Research ethics practitioners may ask questions like "What should the attitude be for researchers who work with human subjects? What rights and responsibilities do researchers have regarding human subjects?" (Mishra, 2021).

In formulating policies for how researchers can work ethically with Big Data, scholars have thus drawn upon existing developments in research ethics. University researchers have a long tradition of establishing and adhering to regulations governing how research involving human and animal subjects occurs. Professors and students must request permission from their university's Human Subjects Review Board before carrying out a research project like a focus group, an interview, or a social science or medical experiment. They must provide documentation attesting to how they will prevent abuses from occurring, how the data will be used, and how participant privacy will be upheld. To secure research grants from both private foundations and government entities, researchers must also provide evidence that they have received permission from the university's Human Subjects Review board to carry out the proposed research.

Specific professional organizations (like the American Psychological Association) also have their own rules and regulations regarding research ethics. Such strict regulations had arisen due to historical abuses, such as when people were made the subjects of medical experimentation without their consent or even their awareness.

But while the rules governing what researchers may and may not do in "live" research regarding people's physical bodies, the regulations regarding using people's data are somewhat newer. They are still evolving, and there is no clear consensus around many issues. It is not always clear whose responsibility it is to safeguard people's rights about their data and for what sorts of restrictions it is reasonable to ask people to place upon themselves and their organizations regarding gathering, sharing, and analyzing this data.

How Do Researchers Use Big Data? Individual Versus Aggregate Data

Big Data, once collected, can be used in two different ways. *Aggregate data* (or data about a group) can identify general patterns in people's behavior. For example, a researcher might look at data about website visits to an outdoor company to formulate a principle such as "most people who buy hiking boots buy them in September." The company might then decide to have a sale in September.

Personal or customized data is data about a particular individual's browsing behavior. In the hiking boot example, individual data might note that someone has visited the outdoor store's website multiple times without buying. That particular customer might then receive a pop-up with a coupon to entice them to buy. As a result of customization, different people might pay different prices for the same hiking boots, depending on whether or not they received the coupon.

The two examples differ because the aggregate data researcher did not cause a particular individual to undertake a particular action. In contrast, in the second example, the researcher arguably interfered in the research process by prompting or nudging the visitor to undertake an action. In general, research ethics for projects involving human subjects (such as those governing a project undertaken at a university) require that a research subject be informed that they are part of a study or project; that the subject gave consent to participate in the study or project, and that the study is designed to minimize any impact upon the human subjects.

In contrast, the person who received the personalized coupon likely did not realize that he was being "experimented upon" in this way, nor did he consent to participate. So, were his rights violated in some way?

This was the question at the heart of the Cambridge Analytica Scandal. In 2016, researchers from the British firm Cambridge Analytica utilized data collected from approximately 87 million Facebook user profiles. The individuals were then shown psychologically tailored advertisements to change their voter preferences for the 2016 US presidential election. As a result of these actions, Cambridge Analytica was investigated by Britain's Information Commissioner's Office and was also banned from Facebook. Critics argued that the company had engaged in data misuse and mass manipulation – since participants did not know that they were being studied or that the company was seeking to elicit a particular reaction from participants. The "study" was also accused of being an attempt at election interference and violating national sovereignty, which dictates that other nations should not interfere in foreign elections.

Big Data thus opens up the door to situations in which data is not merely collected but can be manipulated to influence data producers or data subjects. Using a utilitarian ethical framework, one might argue that if an appropriate social good is sought through such experiments or manipulations, the result is perhaps more important than the specific individual rights to consent and information of those data subjects. For example, one could argue that a firm that created targeted ads to convince people to get vaccinated against a pandemic created a public good – health safety – and that these positive effects of creating that good outweighed any potential harm to individuals. However, others would argue that there is no situation in which it is appropriate to violate the rights of data subjects.

The Power of Big Data

Winston Churchill once said, "where there is great power, there is great responsibility." Today, in many instances, the old statistical techniques (in which one took a sample of people responding to a survey and then generalized to the entire population using statistical inference) are being replaced by new techniques, in which sampling is not necessary because we now have all the data or data about all of the respondents in a population. This possibility of having "all the data" can change how researchers think about causality and the modeling of decisions.

It's possible to speculate about a future in which humans, fueled by "all the data," are omniscient. Could we as humans someday be omniscient – possessing all of the human knowledge and the ability to predict things that haven't happened? There are indicators that we are progressing in this direction.

However, as we saw earlier in this text when considering the issue of privacy, data can be used to empower both those whose ends are benevolent or helpful, and those whose ends are not. The US Department of Justice defines *online radicalization* as "the process by which an individual is introduced to an ideological message and belief system that encourages movement from mainstream beliefs toward extreme views, primarily through the use of online media, including social networks such as Facebook, Twitter, and YouTube" (United States Department of Justice, 2014). Online manipulation techniques, or the "curation" of the content one sees online, can push people to adopt more extreme and polarizing political views or create a volatile emotional climate online, which can spill over into the real world.

Big Data and the Problem of De-Identification

Earlier in this chapter, we referenced the idea of "downstream" uses of data. Data may be collected in one set in one form, but it can be "bundled together" with other data from other sources or used by other researchers studying other problems. One issue that occurs when data is aggregated in this way is that in some circumstances, as more data about users in a geographic area or demographic is married up with other data collected at different times using different means, anonymity may break down. Think of your unique characteristics: If someone knew that a female individual born on a specific date, living in a specific zip code, and attending a particular college had recently received a speeding ticket, how long would it take to figure out that it was you? We are all unique enough that there may be specific activities and identifiers that form a unique pattern.

And your data might accidentally give away personal information that you were not aware you were furnishing. A computer researcher describes how a bank noted an unusually high number of ATM cash withdrawals occurring between midnight and 2:00 a.m. in a specific location and became concerned that perhaps some form of theft was occurring. Upon investigating, the bank found out that the ATM was located near a "red light district" (or place where sex work occurs). The bank found that, as a result, it had a list of which of its customers were likely hiring sex workers. These participants didn't "opt in" to furnishing this data and didn't know it was being collected.

"Sharing Is Caring": Is User Data an Individual or a Collective Good?

Another way of considering individual versus group rights to data is to ask whether data is an individual good that belongs to the user, or whether data is better understood as a collective good to which everyone in society has a claim.

In Dave Eggers's 2013 satirical novel *The Circle*, we see the moral evolution of the twentysomething main character, Mae Holland, after she works at a social media company that resembles Facebook. In one of the book's pivotal scenes, Mae has visited her dying father, and afterward, feeling upset, she seeks to comfort herself by taking a kayak out onto the water. The next day, she is called into her boss's office. Her superiors are upset that she has referenced her kayak experience on social media but that she has not chosen to upload any video of the event (Eggers, 2013).

In the dressing down that Mae receives from her boss, we can identify how Eggers uses all three types of ethical arguments (virtue ethics, practical ethics, and deontological ethics) to argue that Mae should have felt morally obligated to share her data with the larger world and that her decision not to share her data at that moment was selfish. First, her boss reminds her of the organization's motto, "Sharing Is Caring," and asks her if she meant to be the kind of person who kept her experiences to herself, hoarding them up rather than sharing. Next, he asks if it wouldn't have been better to share her experience so that more people could enjoy it. Finally, he references a family member who lives with a disability and cannot go kayaking himself. He argues that this family member might have benefitted from experiencing Mae's kayak adventure vicariously by watching it online. "Imagine this was you," he asks. "If you were unable to go kayaking, wouldn't you want others to share that experience?"

When the novel was written, Mae's boss's language sounded hyperbolic or over the top. It was difficult to imagine a scenario in which people might feel compelled to share their data or to sustain the argument that others might somehow have a claim to your data.

However, today, this moral and ethical argument does not sound nearly so implausible. Today, we often distinguish between two different types of privacy: *information privacy* and **data privacy**. The claim that someone "owns" their data is not always accepted, and those who work with data and information may have different approaches to guaranteeing the safety and security of that data than they do to guarantee an individual's safety and privacy based on different understandings of how the two types of privacy work and are configured.

In *The Circle*, Mae's boss made a utilitarian argument, claiming that data needs to be shared if, by doing so, others can benefit from an experience. Indeed, data sharing for the public good is not new. For example, the apps you use to predict traffic and congestion (like Waze or your GPS) start collecting aggregates (or group) data such as the number of cars on the road, people's commuting patterns, and data collected by toll systems like E-ZPass. In the same way, many education researchers claim the valuable data students generate by participating in online educational platforms like Coursera, Blackboard, and Canvas. Researchers seek to use this data to ask questions like "Is there a correlation between how long students spend online in a learning management system (LMS) and their grades in their courses?" But to whom does this data belong, and do researchers automatically have a claim to it? And what about health data? Indeed, if data about hospitalizations can be used to understand better the progress of disease and the most effective treatments, should such information be shared?

As Klose et al. (2020) note, in the United States, the Family Educational Rights and Privacy Act (FERPA) states that students have control over who may access their educational records. But, as they note, a debate exists regarding what constitutes an educational record. Do they argue that a student's grade on an exam is an educational record and thus protected by FERPA? What about a record of how a student interacted with an LMS over a semester? If data such as how many times students opened a file or retook an exam or how long a student spent on activities were made available to researchers, they might be able to refine better their teaching methods or their methods of presenting information in an online environment. The LMS developers might also use this data to build a better product. But as Klose et al. note, FERPA isn't clear regarding who owns this "click data" and whether FERPA covers it. Furthermore, they note that the sorts of protocols and procedures that might be sufficient for safeguarding user privacy regarding one data point or one educational record in isolation might break down once multiple records and databases are combined (that is, given enough data, it may be possible to work backward to identify a user or a school or a school district).

Box 9.2 Going Deeper: What Are Students' Rights Regarding Their Educational Data?

Do you know what sort of information is contained in your educational records? As a recent study points out, universities have access to a great deal of sensitive data about students and their parents' finances, whether students have sought treatment at a university health center for medical or psychological distress, students' grades, their Social Security numbers, their test scores, and even their police records if they have been disciplined through campus police. How well do universities fulfill their role of safeguarding student information from cybercrimes and attacks, how else might they use this information, and under what circumstances?

In a recent survey, many universities were vulnerable to cyber-attacks according to the standard vulnerability scoring system CVSS. Some universities were found to be using outdated software, which contained vulnerabilities (Risk Recon, 2019).

Regulations Regarding University Data Security

In the United States, the Gramm Leach Bliley Act (GLBA), passed in 1999, requires postsecondary institutions to ensure student financial aid records, information security, and confidentiality. The US Department of Education also recommends that educational institutions comply with the National Institute of Standards and Technology Special Publication 800–171. Universities may encounter additional requirements if they undertake research in cooperation with partners such as the National Institutes of Health, the Department of Defense, or Education. US institutions that enroll students from the European Union, including online education, may also be subject to the European Union's General Data Protection Regulation (Tech Totem, 2021).

Universities Using Your Data for Education Research

Researchers, particularly those in the field of education, also see student data about students and student-generated data (such as records of student actions or interactions in a **learning management system** or LMS such as Blackboard or Canvas) as a valuable source of data for the researcher. Through accessing university data, researchers can ask questions such as how first-generation students avail themselves of university resources in comparison to other students, or how English-language-learner students make academic progress from their first to their final year of study.

But wait! Were you, as a student, aware that researchers were using your data in this way? Did anyone ask you for your permission?

As Jones et al. point out, the rules regarding student data ownership are somewhat murky in the United States. They ask: "Who owns the data used in large-scale analytic processes on college campuses, especially when the data is personally identifiable to the student?" (Jones et al., 2014). In their analysis of the situation, they note that in the United States, the Family Education Rights and Privacy Act (FERPA) protects students' data, which cannot be released without their permission (for example, to a prospective employer). However, it is less clear what protections exist when these records are used at an aggregate level.

Competing Ethical Goals

Here again, we see two competing sets of ethical goals at play. Many university researchers and administrators argue that having access to student-generated data is necessary to serve all of their students best. And, they might add, for universities to be financially competitive in a tight education market, they must offer the highest-quality education to all students. By looking at what works and what doesn't work for student learners, administrators and educators can better assure that students receive appropriate levels of counseling, that they are guided to complete their education within four years, and that students who are at risk and struggling are identified. Privacy risks and concerns are outweighed by the ethical goals of increasing diversity, equity, and access, as well as by the goals of efficiency and effectiveness.

In contrast, students may feel that the university's efficiency and effectiveness goals do not justify using student data without permission, and that the risks to student privacy and the necessity of consent are more important factors to consider.

Bibliography

Jones, K., Thomson, J., & Arnold, K. (2014). *Questions of data ownership on campus.* Retrieved June 14, 2022, from https://papers.ssrn.com/sol3/papers.cfm?abstract_id=2540337.

Risk Recon. (2019). *How universities fail to protect student data.* Retrieved June 1, 2022, from https://blog.riskrecon.com/universities-student-data.

Tech Totem. (2021). *Cybersecurity expectations for higher educational institutions.* Retrieved from https://www.totem.tech/cybersecurity-requirements-for-colleges-and-universities/.

A related ethical issue is whether people can or should be compelled to produce and provide such data in instances where collecting and analyzing collective data is an effective public policy intervention (such as in an instance of disease surveillance). For example, should everyone be required to download a disease surveillance app (which relies on geolocation data) onto their phone to improve the collective data collection and aggregation effort?

What about a situation in which people are utilizing self-driving cars? In 2017, the German Federal Ministry of Transportation argued that "more data equals more safety" in arguing that algorithms would be strengthened if drivers were required to share their car's data with other drivers. In this instance, some analysts have even argued that the data generated by your self-driving car does not properly belong to you but rather to your automobile's manufacturer. The manufacturer must own this data; the argument goes for the company to continue improving the algorithms that govern your automobile's behavior (Schneier, 2015). Zhang has argued that individuals do not have the right to own raw data generated in this way in the same way that they have a right to own the data which they are consciously generating. Furthermore, he suggests that a company might be able to regard the data generated by its vehicles as a sort of "trade secret" that it owns, regardless of who is driving the vehicle. He proposes a future scenario in which perhaps self-driving cars will have a black box data recorder similar to that found on airplanes. In the event of a crash, personnel from the US Transportation Security Administration (TSA) would be able to access the information to address why the crash occurred and hopefully prevent future crashes from occurring.

Zhang suggests that an emerging legal framework governing data ownership might be patterned on the *United States Health Insurance Portability and Accountability Act* (*HIPAA*), regulations governing how your medical information may be shared and stored. In this framework, a hospital might have access to your medical data. Still, clear rules exist about who may look at it and under what conditions, how it must be stored, whether it can be sold or shared with medical researchers, anonymity practices that must be maintained, and so on (Zhang, 2018). However, currently there is no clear regulatory framework governing who is responsible for the cybersecurity of data generated by a self-driving car, whether companies are required to report hacking attempts and to whom, or who may legally access vehicle data and under what conditions. (For example, could your insurance company or law enforcement access this data?)

Data Ownership and Ethical Issues from Storage to Consent to Surveillance

But what specific ethical tasks or responsibilities do those in the field of cybersecurity have when it comes to working with Big Data? Those who work with Big Data face new ethical challenges due to Big Data's nature as an emerging technology. Decisions that may seem merely like engineering decisions – like where and how to store data, categorize data, and determine access levels for data, and how and when to merge data streams – can have real-life repercussions.

Cybersecurity practitioners will encounter ethical issues related to Big Data as they work to provide information assurance. Information and data are valuable assets, and cybersecurity experts help guard this valuable information. The US National Institute on Standards and Technologies defines information assurance as "measures that protect and defend information and information systems by ensuring their availability, integrity, authentication, confidentiality, and non-repudiation." Cybersecurity practitioners work to provide *data integrity* – by making sure that data has not been altered in an unauthorized or accidental manner since it was created, transmitted, and stored. First, practitioners must ensure that sensitive data is stored with appropriate encryption to minimize the chances that such data will be hacked. They are obligated to ensure that the computers used to store and work with this data have up-to-date operating systems, firewalls, and antivirus protection and ensure proper training for everyone who will handle the data. Cybersecurity practitioners are often also concerned with making sure that the facility where data is stored is physically secure – that it has locks and a guard, and that the computers are air-gapped rather than connected to the internet. They should also ensure that the enclave is virtually secure or that secure low-bandwidth connections can reach only the computer systems. Cybersecurity practitioners may also be involved in setting protocols and privileges and determining who may view, work with, or export the data.

The term *data curation* refers to managing data through a data lifecycle – ensuring that it doesn't become corrupted and is not inaccurate. Cybersecurity practitioners provide for non-repudiation by showing the source or origin of data and assuring that someone cannot deny the validity of the data.

Box 9.3 Critical Issues: Data Governance and Data Trusts

What is a *data trust*, and how can it be used to protect users' data online? In the United Kingdom in particular, some analysts have suggested that the best way for users to have a say into how their data is shared, stored, and used is for entities known as data trusts to be established. In a 2017 British government report, *Growing the Artificial Intelligence Industry in the UK*, data trusts were defined as "not a legal entity or institution, but rather a set of relationships underpinned by a repeatable framework, compliant with parties' obligations, to share data in a fair, safe and equitable way" (Rinik, 2020).

Currently, within the European Union, data is regulated according to the General Data Protection Regulation, which provides for top-down data governance. Citizens, or *data subjects*, have certain rights, but the obligation for protecting data is assigned to a data controller. Businesses must appoint a *data controller* who decides how best to make sure that this organization is adhering to the regulations about data protection and sharing. Citizens consent to certain practices, as you do when you agree to terms and conditions when downloading a new app, for example (Reuters, 2018).

What Is a Data Trust?

However, in comparison to the data controller model, the data trust model is a bottom-up model. In a data trust model, users can choose to pool the rights they have over their personal data within the legal framework of a data trust, and data trusts will exist in a variety of formats, including public and private initiatives. In a trust, we are not merely "data subjects." Instead, we give over our data (which we are presumed to own) to a trustee who then acts on our behalf, and who has specific duties and responsibilities. The trustee is thus like a data controller, but the trustee has responsibilities both to society *and* to those people whose data he is responsible for. Trustees can be mandated to consult with the settlors or beneficiaries. The trust is thus like a cooperative in which all have rights and responsibilities.

The notion of a data trust is not completely novel; rather, it rests on certain preexisting understandings about how we as citizens interact with our property in various settings. As citizens we regularly entrust parts of our property to others – for example, we entrust our money to a bank. In a data trust model, then, individuals would be able to "shop around," deciding where to entrust their data and to whom, just as you might decide to which bank to entrust your money.

People would thus enroll in a data trust or even several different data trusts. For example, you might decide to put your medical data into a certain trust, and the trust would then grant rights to that data to various types of researchers. You might have the ability to vote as a member of a trust, and to green-light or red-light certain research initiatives (Delacroix, 2019). In situations where your data is being monetized, you might even have an ability to share in the profits!

Separate trusts might exist for your social media data or your genetic data, as well as your financial data or loyalty card data. And in a future scenario where you perhaps live in a smart city and are continually producing data as you navigate that city, you might participate in a data trust with others who live in the city, deciding who would have access to that data and for what purposes (Rinik, 2020). A data trust's members might also vote, for example, on how their data should be stored and what sorts of anonymization and encryption procedures might be used.

Problems With the Data Trust Model

However, not everyone agrees that data is "property" that can be "vested" in a trust in this way. They view data instead as an intangible asset. To give an analogy, some people believe that data is not like a house that you own, but instead like a river that flows through your property to which you might have rights – but which you do not actually own outright.

Others have identified logistical challenges, such as the difficulties that might be encountered when one wishes to exit a data trust. (Issues regarding the disposition of your data should you die also exist.)

Finally, there are issues to be resolved regarding the liability a data trust manager might encounter if, for example, individuals were harmed in some way by the actions taken by the data trust.

Bibliography

Delacroix, S. A. (2019). Bottom-up data trusts: Disturbing the "one size fits all" approach to data governance. *International Data Privacy Law*, *9*(4), 236–252.

Reuters, T. (2018, May 25). Top five concerns with GDPR compliance. *Thomsonreuters.com*. Retrieved from https://legal.thomsonreuters.com/en/insights/articles/top-five-concerns-gdpr-copmlicance.

Rinik, C. (2020). Data trusts: More data than trust? The perspective of the data subject in the face of a growing problem. *International Review of Law, Computers and Technology*, *34*(1), 342–363.

Cybersecurity practitioners provide *information assurance* in compliance with the specific rules of their nation or geographic area. However, as we have seen throughout this text, cybersecurity practitioners often have multiple clients and stakeholders with multiple competing interests. For example, a CEO may want to sell data about clients to increase the gains for shareholders even if data producers are unaware or don't wish to have their data shared in this way (Schou & Hernandez, 2014).

Thus, information assurance strategies and practices may change depending on the nature of the task. The need to safeguard information may change, and the sensitivity level of information may change over time. The number of users who require access to data may change, and regulations governing data may change. Thus, cybersecurity practitioners will need to be flexible.

In addition, cybersecurity practitioners encounter ethical issues, including when one is obligated to report vulnerabilities and attacks on extensive data systems, when one is obligated to respond to vulnerabilities and attacks, and how one can best predict and prepare for new vulnerabilities and methods of attack.

Cybersecurity practitioners can thus harness the power of Big Data to predict what sorts of vulnerabilities and threats against their systems are likely to emerge. The new field of cybersecurity vulnerabilities analysis uses web scraping and other techniques (including crawling through cybersecurity blogs and websites, Reddit posts, and RDF site summary feeds) to collect datasets about web breaches to correct and forecast future threats. Information about web breaches is shared to the *Common Vulnerabilities and Exposures (CVE) database* and the United States' National Vulnerability Database. Researchers can analyze these resources to find trends and tendencies of attack and see the relationships between attackers (Tang, Azazab, & Luo, n.d.); in a sense, what these researchers are doing is using Big Data to protect Big Data.

Box 9.4 Going Deeper: Ethics of Cryptocurrency

As we learned in Chapter 3 (Box 3.4, "Tech Talk: Cryptohacking and Cryptojacking"), **cryptocurrencies** are a nontraditional means of carrying out financial transactions. While traditional currencies are issued by a central bank and administered through a nation's monetary policy (including setting interest rates and exchange rates), cryptocurrencies represent a decentralized form of exchange that is not subject to a central set of monetary policies. As referenced in Box 3.4, cryptocurrencies are "mined," not issued (through solving significant technical problems on computer systems configured for these purposes). They are administered and registered via a *decentralized ledger technology (DLT) system on a peer-to-peer network*. Transactions thus exist on the blockchain and are cleared through the DLT system rather than through a central bank like traditional financial transactions. In addition, cryptocurrencies exist only virtually – they are not tied to a specific tangible good (such as currencies linked to the gold standard) and do not exist in paper form.

For this reason, in the United States, cryptocurrency is not regulated by the Federal Reserve but rather by the Securities and Exchange Commission (SEC). This body typically regulates securities such as stocks and bonds (Chaffee, 2019). And the SEC does not make monetary policy the same way that the Federal Reserve does but merely enforces existing regulations regarding securities.

What Is the Upside of the Cryptocurrency Exchange Market?

Cryptocurrencies are sometimes described as existing and operating in a more democratic form than traditional currencies. While a traditional bank may require patrons to maintain a minimum balance or provide identification documents to open an account, anyone may trade in cryptocurrencies. In this way, cryptocurrency might be described as a more equitable economic system with few barriers to access and in which anyone may participate. (That is, it may provide a better place for someone undocumented to save their funds than a traditional bank, which may require legal documentation to open an account.) For this reason, participating in the cryptocurrency market may be particularly attractive to those in the developing world who are closed out of traditional economic systems since they may have a low or erratic income, may lack proper documentation, or may receive their income through quasi-legal arrangements (such as working "off the books" providing maintenance or childcare, for example).

In addition, some ethicists argue that cryptocurrency provides a good way for citizens to "go around" central banks in a nation where the taxation burden against ordinary systems may seem unjust or unfair (Dierkesmeier & Seele, 2018). Indeed, migrants who work abroad and send funds home to their family members may use cryptocurrencies to send money directly without declaring the funds or paying taxes.

On a larger scale, as Scharding points out, if people are fundamentally opposed to capitalism as an economic institution, then refusing to "play the game" by declaring one's assets and paying taxes may be seen as a weapon that can be used to fight back against the capitalist system. In addition, those who adopt a libertarian (or "small and limited government") orientation towards politics may also believe that carrying out one's financial transactions through a system that does not involve the government is inherently the ethical or moral thing to do – since it deprives a central authority both of revenue and of information regarding how, why and when people spend their money (Scharding, 2019). Finally, some radical economists believe that the growth of cryptocurrency could lead to new social formations, giving cryptocurrency the ability to fundamentally remake our societies in new and more equitable ways through speaking back to globalism (Doria, 2020).

What Is the Downside of the Cryptocurrency Exchange Market?

As noted earlier, cryptocurrencies and transactions on the cryptocurrency market are not the subjects of government regulation in the same way traditional currency transactions are. In the wake of the 9/11 terrorist attacks against the United States, many nations instituted strict monitoring systems to ensure that people were not using traditional banking systems for money laundering and terrorist finance. People engaging in significant financial transactions at a traditional bank may be asked to provide identification, observe a waiting period before a transaction clears, and observe specific limits on the size of transactions they may carry out. In contrast, cryptocurrency transactions can be carried out anonymously without providing identification and, at least at present, are not the subject of the same government regulations.

For this reason, law enforcement agencies have warned that cryptocurrencies can be used to carry out illegal and unethical financial transactions and that practices stem from cryptocurrency. In recent years, many ransomware attacks have included a demand for payment in cryptocurrency. In the United States, the Department of the Treasury has listed several cryptocurrencies on its list of entities which US firms are banned from engaging with the *Specially Designated Nationals and Blocked Entities (SDN) list* (Shook, 2021).

Cryptocurrency markets are also much more volatile than traditional means of exchange. In a traditional banking system, the government may step in to, for example, adjust interest rates or enact policies to stem inflation. Since no similar mechanism exists in a cryptocurrency market, some analysts argue that it is unjust to encourage people with limited savings and a lower tolerance for risk to invest in cryptocurrency (Dierkesmeier & Seele, 2018) – even though the system's lower barriers to entry may make it seem particularly attractive to just such an audience. In this way, the risks created by the crypto market may be unfairly distributed, landing particularly harshly on the most vulnerable.

For this reason, in September 2021, China's Central Bank announced that "virtual currency-related business activities are illegal financial opportunities . . . since it seriously endangers the safety of people's assets" (BBC News, 2021).

Furthermore, some economists argue that the growth of cryptocurrencies as a type of security increases the risk to the securities market (or stock market). They worry that a significant collapse in the cryptocurrency market could imperil other investments that citizens have made, affecting groups like the elderly who may have their retirement savings invested in the stock market (Chaffee, 2019).

Finally, many politicians and economists believe that the ability to administer a nation's economic system is a fundamental right that should belong to a nation's leaders (and, by extension, to citizens who elect these leaders). They argue that leaders should be able to protect people and their investments by making monetary policies. One of the reasons that citizens trust their leaders (and our system of democracy) is because their leaders can do this. In this way, they argue, the growth of unregulated cryptocurrencies undermines people's economic rights to live in a country with a stable government and currency and their right to have leaders who can guarantee that stability (Scharding, 2019).

In the Future

As Swartz points out, the fundamental ethical issue that arises in considering the growth of cryptocurrencies is broad: What is money, and what is it for? Should money be regarded first and foremost as a tool, and if so, who should wield that tool, and for what purposes? Should governments make monetary policy to achieve specific goals (such as creating a stable field for people to save and invest) for the good of society, or should individuals be able to make their own unregulated decisions to achieve their own personal and private goals? These issues will interest cybersecurity practitioners and experts as they evolve in the coming years.

Bibliography

BBC News. (2021, September 24). China declares all crypto-currency transactions illegal. *BBC News*.

Chaffee, E. C. (2019). The heavy burden of thin regulation: Lessons learned from the SEC's regulation of cryptocurrencies. *Mercer Law Review*, *70*(1). Retrieved June 1, 2022, from https://papers.ssrn.com/sol3/papers.cfm?abstract_id=3371667.

Dierkesmeier, C., & Seele, P. (2018). Cryptocurrencies and business ethics. *Journal of Business Ethics*, *152*(1), 1–14.

Doria, L. (2020). Cryptocurrencies: Economic and social dimensions. *Parte ci Pazione e Conflitto*, 384–408.

Scharding, T. (2019). National currency, world currency, cryptocurrency: A Fichtean approach to the ethics of bitcoin. *Business and Society Review*, *124*(2), 219–238.

Shook, H. A. (2021). *To pay or not to pay: What new regulatory activity means for ransomware vicitims*. Retrieved June 3, 2022, from www.jdsupra.com/legalnews/to-pay-or-not-to-pay-what-new-4721511/.

Swartz, L. (2018). What was bitcoin? What will it be? *Cultural Studies*, *32*(4), 623–650.

The Problem of Competing Ethical Values

As noted earlier in this text, it may be difficult in some instances to define the one best solution to an ethical problem. Data engineers will encounter ethical issues as they attempt to balance the needs and desires of the data users with other concerns, such as storing data in fiscally responsible ways since data storage can be costly. They may also encounter situations in which there are multiple sets of stakeholders with competing expectations and desires about the data. For example, encrypting data or applying privacy protocols like anonymization may make it easier to secure the confidentiality of those whose (medical or educational) data it is, but doing so may be more expensive or make the data less useful for other features. For example, it may now be more difficult to combine the data with other data streams.

In addition, a client might wish to sell or share data with another corporation – but every time data is moved or shared between organizations, the risk of a breach increases. Different organizations may have more or less expertise in dealing with data, different protocols and levels of access to this data, and different personnel responsible for this data. They may also have different data governance policies. **Data governance** refers to setting internal standards (or data policy) for gathering, storing, processing, and destroying data. A data governance policy specifies who can access what kinds of data and the data types subject to these rules. Thus, for example, a university might have a data governance policy specifying what information a faculty member might have access to in student records versus the sorts of records that someone in financial aid might access; here, a data engineer might feel ethically obligated to preserve the privacy of those whose data it is and feel obligated to the needs of a corporation's shareholders and its leaders.

Those who work with data will frequently have to balance and make trade-offs between reducing and safeguarding against risk, preserving the utility of data, treating data most efficiently, maximizing safety, and ensuring data subjects' privacy.

CHAPTER SUMMARY

- Data exists in various forms – and raw data can be transformed in various ways. New technological developments allow us to transform and aggregate data to generate new insights about our social world.
- Engineers exercise power by making decisions about data governance – including defining who should have access to data and the sorts of restrictions that might be placed on data access.
- Ethicists disagree about whether data is best understood as a collective good – where sharing data can lead to solutions to social problems – or an individual good – since people's right to control how their data is used and by whom should be safeguarded.
- Data architects and analysts often face a dilemma of competing values – balancing needs like efficiency and accuracy against needs like privacy and security – in configuring a data architecture.

DISCUSSION QUESTIONS

1 **The Future of Big Data**

Watch the TED Talk by Zeynep Tufekci, November 12, 2017, "We're Building a Dystopia." Available at www.youtube.com/watch?v=iFTWM7HV2UI (accessed June 21, 2022).

- What are some of the social problems that Tufekci believes might arise from algorithms that "steer" users to certain types of information?

- She notes that what designers want to achieve – selling more ads – differs from the users' goals (such as finding accurate information about current events or a medical problem). She feels that users are *owed* a more transparent system and where they are not subject to decision-making "behind the scenes" of which they are unaware. Do you agree that corporations have an ethical responsibility to consider what users want?
- How might user-centered design principles be implemented to correct a situation as she describes.

2 **When Is AI Too Powerful?**
David and Patterson note that something may merely feel creepy (i.e., how did Alexa know that I am interested in this news story or that I like this type of music?), but that creepy feeling may be a clue that some form of ethical decision point exists (David, 2012). Think to yourself:

- Did you ever have an encounter with technology that felt "creepy" to you, or where you asked yourself, "How did my device know that about me? Why did it suggest that product or movie, etc.?"
- What ethical decision-point was represented at that moment?

3 **A Bill of Rights for Data Producers**
What might a bill of rights look like in relation to your data?

o What rights should citizens have in relation to their data?
o Should they have a right to know when it is being collected? A right to know when it is being shared or a right to have it erased or deleted?

- Is it technologically feasible to have such a bill of rights?
- What is the likelihood of achieving international consensus on an international bill of rights in relation to data?

4 **Trade-Offs and Competing Values**
In thinking about trade-offs involved in data sharing scenarios, which rights do you see data subjects as having, which rights do you see the public as having, and which rights do you think company shareholders have? For example, think about the collection of people's data to create a vaccine.

RECOMMENDED RESOURCES

Jones, K., Thomson, J., & Arnold, K. (2014, August 25). Questions of data ownership on campus. *Educause Review*. Retrieved from https://er.educause.edu/articles/2014/8/questoins-of-data-ownership-on-campus.

3Blue1Brown. (2017, July 7). *But how does bitcoin actually work?* Retrieved June 21, 2022, from www.youtube.com/watch?v=bBC-nXj3Ng4.

World Wide Web Foundation. (2018). *Open data barometer – leaders edition*. Washington, DC: World Wide Web Foundation. Retrieved from opendatainitiative.github.io

References

Chaffee, E. C. (2019). The heavy burden of thin regulation: Lessons learned from the SEC's regulation of cryptocurrencies. *Mercer Law Review*, *70*(03–035), 615–640.

Cox, A., & Verbaan, E. (2018). *Exploring research data management*. London: University College.

David, K. (2012). *Ethics of big data*. Sebastopol, CA: O'Reilly.

Eggers, D. (2013). *The circle*. New York: Knopf.

Garfinkel, S. (2018, January). Privacy and security concerns: When social scientists work with administrative and operational data. *American Academy of Political and Social Science*, 83–101.

Jurkiewicz, C. (2018). Big data, big concerns: Ethics in the digital age. *Public Integrity*, *20*(1), 546–559.

Klose, M., Vasvi, D., Song, Y., & Gehringer, E. (2020). EDM and privacy: Ethics and legalities of data collection, usage and storage. In *13th annual conference on educational data mining*. Virtual: International Educational Data Mining Society. Retrieved from https://eric.ed.gov/id=ED607820.

McCormack, A. (2016). Downstream consent: A better legal framework for big data. *Journal of Information Rights, Policy and Practice*, *1*(1). http://doi.org/10.21039/irpandp.v1i1.9.

Mishra, N. (2021). International trade law meets data ethics: A brave new world. *New York University Journal of International Law and Politics*, *53*(2), 305–347.

Rinik, C. (2020). Data trusts: More data than trust? The perspective of the data subject in the face of a growing problem. *International Review of Law, Computers and Technology*, *34*(3), 342–363.

Schneier, B. (2015). *Surveillance is the business model of the internet*. Retrieved from https://www.schneier.com/news/archives/2017/07/surveillance_is_the_business_model_of_the_internet_html.

Schou, C., & Hernandez, S. (2014). *Information assurance handbook: Effective computer security and risk management strategies*. Columbus, OH: McGraw-Hill.

Tang, M. J., Azazab, M., & Luo, Y. (n.d.). Big data for cybersecurity: Vulnerability disclosure trends and dependencies. *IEEE Transactions on BIg Data*, *5*(320), 635–85.

United States Department of Justice. (2014). *Online radicalization to violent extremism*. Washington, DC: United States Department of Justice.

Zhang, S. (2018). Who owns the data generated by your smart car? *Harvard Journal of Law and Technology*. Retrieved from https://thefreelibrary.com/WHO+OWNS+THE+DATA+GENERATED+BY+YOUR+SMART+CAR%3F-90575356483.

10 Military Aspects of Cybersecurity Ethics

LEARNING OBJECTIVES

At the end of this chapter, students will be able to:

1. Point to emerging ethical issues in the conflict in cyberspace, including issues related to agency, autonomy, and responsibility
2. Describe the cognitive domain as a new challenge in warfighting, and its relationship to disinformation strategy and tactics
3. Describe how existing laws and norms governing the conduct of war may be more difficult to apply in situations of AI-enabled warfare

As we consider the ethical problem of cyberwarfare in this chapter, we can consider five real-life situations involving the use of cyberwarfare or response to cyberwarfare by a government, corporation, or individual.

- The Ukraine conflict began in February 2022 with Russia's invasion of Ukraine and has been described as the "world's first open-source war." Information about troop movements, planned invasions, and the state of both sides' military readiness and their equipment has been gleaned using technologies like iPhone tracking, satellite images, and the use of location-based services by combatants. Decisive military victories have been attributed to the availability of open-source information, much of it provided by enemy combatants who were often unaware of the information they were making available to their adversaries (Puiu, 2022).
- In the United States, many of the individuals who came to Washington, D.C., on January 6, 2021, to participate in an insurrection, fueled by their beliefs that Donald Trump had been a victim of "election stealing" in the 2020 US presidential election, had been radicalized due to information they encountered on social media. Many analysts suggest that the profoundly polarizing, highly divisive information made available to citizens online was seeded as part of a Russian disinformation campaign to undermine US election legitimacy and credibility (Srinivasan, 2021).
- In 2016, as British citizens voted in a referendum to determine whether or not the United Kingdom would continue to be part of the European Union or whether it would engage in the act of "Brexit," over 150,000 Russian-language Twitter accounts posted English-language messages urging Britons to vote to leave the European Union (Kirkpatrick, 2017).

DOI:10.4324/9781003248828-13

- Specialists in the field of energy security note that attacks on power grids have become an increasingly common type of warfare – carried out by adversary nations and groups like terrorists. An attack on an energy grid could cause hundreds of thousands of people to lose power and have repercussions in areas as diverse as health security, food security, and the transportation sector (Chester, 2020).
- Shortly after the Russian invasion of Ukraine, policymakers began suggesting new types of sanctions that might be used against Russia. Among policy measures suggested was an initiative to cut Russia off from the worldwide internet by eliminating the ".ru" address node from the internet's configuration tree. In addition, Russia has been cut off from accepting international payments as the SWIFT codes for Russian banks are no longer valid.

What do these five incidents tell us about cyberwarfare and the need for cybersecurity? First, these incidents indicate how traditional geographic borders and barriers – like rivers, oceans, and border control checkpoints – are no longer sufficient to defend a nation's internal territory against foreign interference. Today, national sovereignty – or the ability of a state to exercise control over its territory and the functions taking place in that territory – is no longer absolute. Foreign actors can, through the internet, interfere in domestic policy referendums and national elections and target civilian structures that form part of critical infrastructure.

At the same time, these incidents show how warfare between nations and adversarial actors is often carried out not only by professional soldiers, but also through the actions of private corporations and private citizens who may engage in hacking, disinformation activities, or intelligence gathering. International organizations like ICANN (Internet Corporation for Assigned Names and Numbers) and SWIFT may also play a role in situations of conflict. These events also illustrate how civilians can be targeted and mobilized in an ongoing military conflict between adversarial actors.

These incidents also show how every nation is dependent on and interdependent with international technological structures – like the international banking system and the internet. Threatening to cut off a nation's access to international technology infrastructure can be as powerful a weapon as a conventionally armed attack.

Finally, these incidents show how information and data are everywhere – leaking as "data exhaust" from vehicles and warfighters as they engage in combat, and available to both civilian and military observers who seek to target their enemy's personnel and materials.

These incidents show that today, there is often no clear defining line between peacetime and wartime, civilians and military warfighters, and domestic and international borders. There is also frequently no clear distinction between kinetic (conventional) and cyberwarfare, as cyber warfare can be used by a nation to augment its conventional activities through sowing societal confusion before an invasion, shutting down communications in a nation before the invasion, or collecting information (intelligence preparation of the battlefield) before an invasion. Cyberwarfare can thus augment kinetic warfare, while kinetic warfare might also be used to augment cyberwarfare.

Each of these eroding distinctions presents new and challenging tasks for those who work in the field of cybersecurity. At the same time, the erosion of these distinctions means that many of the traditional ethical understandings – which have allowed people to engage in armed conflict within the bounds of civility through establishing clear limits to warfare – are no longer valid and enforceable. Many of our traditional understandings about what warfare is, how it works, and what the ethical limits are no longer seem to hold. At the same time, other facets of warfare, such as how confusion or the "fog of war" often prevails during moments of conflict, still hold and may be more salient than ever. In this new terrain, the question remains: what sorts of new norms can and will evolve to govern the conduct of warfare that depends – more than ever – on technology?

This new terrain is essential for military or civilian cybersecurity professionals to understand as they consider what constitutes ethical actions during cyberwarfare and where the ethical limits may be.

What Is Cyberwarfare?

In the last edition of this book, we described a new and emerging type of warfare known as "cyberwarfare." *Cyberwarfare* has been defined as:

> Warfare in space [which] includes defending information and computer networks, deterring information attacks, and denying an adversary's ability to do the same. It can include offensive information operations mounted against an adversary.
>
> (Hildreth, 2001, p. 3)

Cyberwarfare is thus regarded as part of a more extensive set of operations known as warfare, with "cyber" referring to the fact that specific actions are thus carried out through the use of cyberweapons or cyber activity. One can conduct cyberwarfare independently or as part of a larger strategy in which a nation uses both cyber and conventional weapons and tactics to achieve a target. In some instances, cyberwarfare is seen as a *force multiplier*. A nation uses cyber tactics – such as taking down an enemy's communications infrastructure – to install disorder within the adversary's military or the general population, thus facilitating traditional conventional (or kinetic) attacks.

Attacks using cyber means can be regarded as "acts of war" in some circumstances. In the guidance developed for military officers, decision-makers, and military personnel, an act of war is defined as:

> An action by one country against another to provoke a *war* or an action that occurs during a declared *war* or armed conflict between military forces of any origin.
>
> (USlegal.com, emphasis added)

Today, most analysts do not choose to speak of "cyberwar" as a specific type of war. Instead, cyberwarfare activities, including computer network attacks, malware attacks, and take-downs, are integrated into land, air, and naval warfare. All types of warfare today depend on cyber capabilities, and cyber weapons are used for both offensive and defensive purposes, often augmenting other traditional forms of warfare. In addition, cyberwarfare activities are used for intelligence preparation on the battlefield, and "information warfare" in both its defensive and offensive variants is used for command, control, and communication during all phases of warfighting.

What's Different About "Cyberwar"?

Ten years ago, planners were concerned about how a nation's reliance on cyber capabilities would change the shape and meaning of warfare. Although nations have gone to war for thousands of years, the integration of cyber capabilities means that wars today take place faster, over greater distances, and often include a hybrid of commercial and military actors working together to secure power grids, protect and defend a transportation infrastructure as well as to create and share intelligence.

Because this new type of warfare often does not take place within a clearly geographically defined battlefield (but instead may include individuals overseeing an activity like drone strikes from a location hundreds or even thousands of miles away), and because the line between civilian and military assets is much harder to draw (if, for example, the US military headquarters at the Pentagon hosts its information assets in the civilian Amazon Cloud), the temporal distinction between wartime and peacetime is also often not as clearly drawn. Analysts increasingly speak of "hybrid wars" and "grey zone conflicts' to denote situations in which countries may engage in conflict (including the carrying out of cyber-attacks) without necessarily escalating to a kinetic conflict (one using conventional weapons which may result in physical casualties rather than the destruction of mere data). They may engage in hostilities just short of war, seeking to achieve territorial gains in a situation where an adversary may be reluctant to widen the conflict into a full-scale war.

At the same time, cyber-attacks on civilian infrastructure – including critical infrastructure such as electrical power grids, fuel pipelines, and communications structures – have led to a blurring of boundaries between economic attacks and traditional warfare attacks. Here we can consider an event like the May 2021 ransomware attack on the Colonial Pipeline, which carries 45 percent of the fuel used by citizens on the US East Coast. Pascucci and Sanger (2021) described this scenario as a cross between national security and criminal events. Those who participate in such events may be criminals working independently or individuals engaged in state-sponsored crimes or even terrorism. In this way, events today also represent a blurring of traditional warfare and criminal activity.

Such events illustrate how new and emerging technologies have reshaped or even begun eroding many of our long-accepted ethical and moral understandings about how wars should be conducted and the ethical and moral justifications for the war. In the last edition of this book, we concentrated in great depth on how the ethical framework known as just warfare did and not apply to rulemaking for the conduct of ethical war using cyber components and capabilities. That is, we asked if one could indeed conduct a just war in cyberspace.

As cyber capabilities now form merely one of several types of capabilities that may be brought to bear in a conflict, many analysts have concluded that it doesn't make sense to ask about "cyberwar" as a particular type of conflict or to suggest that wars fought using cyber capabilities should be judged according to different ethics or rules of engagement. In this sense, some believe that the debate about "just war ethics in cyberspace" has been concluded. Rather, as Eaton has stated: Cyberweapons are weapons, and whatever law applies to conventional weapons equally applies to cyberweapons. Cyber is primarily seen as one of many types of capabilities that may be involved in the conflict. With the publication of the NATO-sponsored Tallinn Manual in 2013, which argued that most aspects of international humanitarian and customary international law applied to cyber conflicts and conflicts fought in cyberspace, one can argue that a consensus has been achieved. (Those wishing to learn more about just war principles and their applicability to cyberspace and cyberwar may find additional resources in the Suggested Resources section at the end of this chapter.)

The Emerging Issue of Artificial Intelligence in Military Activities

For this reason, in this chapter, we will concentrate not on just war and the ethics of cyber conflict but instead consider two new and emerging topics in the field of military ethics that are of particular utility to those involved in cybersecurity. First, planners and ethicists are increasingly concerned with how machine learning, algorithms, and artificial intelligence are being integrated as tools useful for war planning and warfighting – including the possibility that someday humans will fight alongside autonomous weapons as part of a human–machine interface, as well as the possibility that fully autonomous weapons systems (AWS) may prepare for, plan, and conduct military conflicts without human oversight or intervention in war planning, warfighting, and autonomous weapons.

Therefore, in this chapter, we will define autonomous weapons and the levels of autonomy. We will then consider ethical arguments for and against using AWS in warfare and how relying on AWS can change the meaning of warfare itself. Here, we consider whether it is necessary to establish ethical limits defining how artificial intelligence and artificially intelligent agents should and should not be involved in military combat. In the future, should AI entities plan and decide military strategies, including making such decisions and how and when human forces are deployed? What are the pros and cons of relying on AI in this way? (Ventre, 2020).

Next, we will consider a new domain of warfare that has recently been defined by both US and NATO military planners. The "cognitive domain" of warfighting refers to how warfighters at all levels – from intelligence experts to planners to those engaged in combat – interface with the information resources available. As noted earlier in this text, artificial intelligence can be used for ethical (benevolent) and non-ethical (malevolent) reasons. In particular, artificial intelligence can create new and consequential types of disinformation and thus can be used to carry out psychological warfare.

Psychological warfare is defined as operations involving the planned use of propaganda and other psychological operations to influence the opinions, emotions, attitudes, and behavior of opposition groups (Rand Corporation, n.d.). Artificial intelligence can be used to augment psychological operations through creating, enhancing, and altering images and videos and manipulating information and perceptions. Again, we discuss the ethical issues associated with military operations in this new domain.

What Are Autonomous Weapons?

The word *autonomous* is from the Greek *autonomous* ("independent"), which is from *autos* ("self") plus *nomos* ("law"). Simplistically, autonomous systems act independently by using artificial intelligence to make independent decisions when faced with unanticipated scenarios (Dowdy, 2018).

However, most analysts (and developers) today distinguish between levels of autonomy when talking about autonomous systems. Sanchez-Herrero (n.d.) distinguishes between three types of autonomous systems. She begins with the most basic or "least autonomous" programs – traditional logic and rules-based systems that can carry out basic everyday tasks like alphabetizing lists and sorting information into categories. She then moves up to more sophisticated algorithms, which allow technology to solve problems that may resemble thinking or mimic human cognition. These systems rest on "general Artificial Intelligence," which mirrors the activities of a human brain.

Finally, she refers to those activities that go beyond human cognition in situations where computers can train, learn, improve themselves, and improve upon existing practices. Such systems, sometimes called high-level or superintelligence artificial intelligence, may involve utilizing more data or variables than a human can track, or working faster than any human could calculate. With the advent of neural nets and quantum computing, many technology analysts anticipate that soon we may experience a "hard takeoff" in which humans are increasingly "outgunned" by computers with skills and abilities that we will never possess. We speak of ceding authority or agency to an artificially intelligent agent and entrusting activities with humans providing minimal or no oversight.

There are many reasons to be optimistic about the potential benefits that autonomous artificial intelligence could produce. AI has the potential to revolutionize how we practice our occupations and make new knowledge in many fields from education, to health care, to city planning, to finance, to the legal sphere, to issues of justice and equity. But it also raises ethical issues.

For this reason, military planners and leaders have begun to think about how autonomous weapons systems can and should be integrated into our fighting forces – and the limits that we might wish to place on such activities. Most robot weapons today are semi-autonomous or "human-in-the-loop" systems. An algorithm might gather data about possible targets in these systems and suggest weapons trajectories. However, the final command to deploy a weapon would still require human intervention.

However, fully autonomous systems are emerging, like the US Navy's Phalanx close-in weapon system. Some states have begun to call for an international preemptive ban on developing and deploying fully autonomous weapons in combat (Solovyeva & Hynek, 2018, p. 171). And in 2016, the US Secretary of Defense Ash Carter stated that the Department of Defense would not employ AI-enabled weapons (lethal autonomous weapons without a human in a loop) unless another state did so first (Baker, 2020).

What Issues Do Autonomous Weapons Pose?

In their work, Solovyeva and Hynek (2018) raise six ethical dilemmas that our shift towards dependence on autonomous weapons systems today raises. These dilemmas range in scope – from more significant ethical issues (such as "how does relying on this technology serve to change the moral and ethical import of war itself") to more prosaic queries about the capabilities of the technology itself, as to how we as humans might be changed or shaped by interacting with and relying on this technology.

Some dilemmas are best expressed by applying a utilitarian ethical framework, while others require adopting a deontological framework. In many cases, one can make both an ethical argument against the use of autonomous weapons and one for the use of autonomous weapons (suggesting that they are more moral or ethical than the alternatives), using the same evidence and argumentation paradigm.

A Deontological View

At its base, the deontological framework emphasizes the golden rule ("How would I feel if this rule which I support was then applied to me?"), and is thus about reciprocity. This lens assumes that to understand the full moral import of my actions and to act morally;

I need to practice reciprocity – to imagine myself acting from the viewpoint of the other person or entity affected by my actions.

But, as noted earlier in our discussion of artificial intelligence, an algorithm is not capable of self-regard. Therefore, it cannot practice empathy, "seeing itself as the other" or seeing the other. This lack of self-regard and other-regard is not ethically or morally significant for many tasks we might depend on. If I acquire a Roomba (robot vacuum cleaner) to clean the floors in my house, it does not matter whether the Roomba can "regard" the dirt, nor does it matter that the Roomba is aware that it is removing the dirt from my floors.

However, when an autonomous machine's actions affect humans subject to decision-making, this lack of self and other-regard becomes troubling. One example of this dilemma is illustrated by an event in 2016. Microsoft proudly debuted an AI-enabled chatbot called TayTweets and assigned its own Twitter account (@TayandYou). The program was designed to learn from interacting with other users online and become more sophisticated as it did so. After only a day or two online, @TayandYou was taken offline – by its human handlers – when it began sharing racist and sexist tweets. The bot could not understand the offense it caused to others by sharing these sentiments. (And even if someday a better Twitterbot were created with built-in safeguards against sharing such information, the bot still wouldn't understand that it had offended, nor would it feel guilty for doing so.)

In the same way that an AI-enabled Twitter bot would not understand that it had offended viewers, an autonomous weapon would not understand that it had taken a human's life, nor would it feel sadness or guilt. For this reason, Peter Asaro, a leading thinker in the movement International Commission for Robot Arms Control (ICRAC), believes that autonomous weapons that can take a human life should be banned. He writes that "it is immoral to kill without human reason, judgment, and compassion. Outsourcing lethal decisions to machines may automatically mean a regress in ethics and morality. Thus, it should be illegal" (quoted in Solovyeva & Hynek, 2018, p. 172).

In addition, some ethicists believe that allowing an AWS to kill humans is dishonorable and not in keeping with the ethics of warfare. Michael Walzer has famously written about war between professional military members as an act of "reciprocal killing." In his view and that of others, what makes war honorable is that the soldier acts bravely, placing his physical safety on the line in the service of his nation and often in service of specific ideas, such as human rights or democracy. In any conflict, he argues, either party runs the risk that he will be killed. He contrasts professional military conflicts with other types of warfare, such as terrorism, in which one might carry out an act against civilians with minimal risk to oneself through, for example, placing a bomb in a schoolyard (Walzer, 2016). Using this logic, one can argue that a soldier who acts to kill others from the safety of a keyboard, or a situation in which the algorithm acts independently to take the life of another, might be viewed as having committed an act of moral cowardice (Manjikian, 2014). The taking of life by a machine, therefore, can also be seen as creating an ethical breach for the human in the loop (who does not present his own body at equal risk in the conflict) and arguably presents an ethical breach for the nation which deploys resources in this way – since the nation also does not put its citizens in harm's way in service of the nation and of ideals.

Machines and the Problems of Attribution and Responsibility

One final ethical issue to consider in our discussion of cyberwarfare ethics is the growing use of automated programs and weapons in cyber hostilities. Miller et al. (2015) note that today, cyber hostilities may be carried out through "soft bots," programs that can carry out activities on someone's behalf. Bots could be deployed to carry out a Dedicated Denial of Service (DDoS) attack. An entity like the Pentagon could find its computers shut down as they are contacted by thousands of computers worldwide, all attempting to reach them simultaneously. And computers could be programmed to automatically "hack back" if they sense that they are being accessed as part of a cyber-attack by a hostile entity in a military protocol known as Active Cyber Defense (Miller et al., 2015)

In Chapter 2, we noted that all three of the ethical models – virtue ethics, utilitarian ethics, and deontological ethics – were created based on certain assumptions, including the assumption that there is an identifiable decision-maker making the moral decision and that the decision-maker was aware that they were doing so. An individual human was ultimately responsible for deploying the virus, engaging in the social engineering activity, or attacking the critical infrastructure. But we see in this chapter that this assumption may not hold when we begin theorizing about cyberwarfare's ethics.

The use of automated technologies in cyberwar fighting raises several ethical issues. First, it raises the issue of deception. Analysts like Miller et al. suggest that it is unethical to dupe an opponent who might believe that he is interacting with a human when he might not. Here, we can use the deontological lens to suggest that such a practice violates the principle of reciprocity. If I would not like to find myself in a situation where I was duped, believing that I was battling with a human when I was battling with a machine, then my opponent would similarly oppose such a practice. Furthermore, traditional military codes of conduct on principles developed in the Middle Ages were called chivalry. Combat is ethical when it is reciprocal, and when both sides are equally likely to be injured or killed in hostilities. It is the willingness to place one's self at risk that renders combat honorable. It is a meeting of equals between two skilled warriors. Thus, some analysts have suggested something cowardly about "hiding behind the technology" by relying on automated warfighting programs and devices. They question how it is possible to call warfare honorable or ethical when it no longer represents a process of joint injury (Manjikian, 2014).

Next, reliance upon algorithms raises legal and moral accountability issues when combat engages. The analyst Luciano Floridi distinguishes between moral accountability and moral responsibility. He notes that a nonhuman entity – like a corporation – can be held morally accountable. For example, one could criticize a company that produces a dangerous product that injures people. However, he argues that the corporation could not be seen as morally responsible since only humans are morally responsible (Floridi, 2013, p. 134). In considering who, then, is morally responsible for the actions of a robot or a drone, one might instead choose to give that role to the person who developed the technology, the person who deployed the technology, the person who gave the order to deploy the technology, or the person assigned to be the robot's "minder." However, Matthias (2004) calls our attention to a responsibility gap – or the possibility that a situation could develop in which the robot is not considered morally responsible for its decisions, but neither is the human operator. She argues that there may be situations where a machine can act autonomously or independently change its behavior during operation in response to changed circumstances, thus creating a vacuum in responsibility.

And Noorman (2014) notes that we tend to have honest conversations that assume that one actor makes moral decisions when most of us exist in webs of social ties. We may carry out tasks with others, and where our actions are dependent upon the actions of others. Thus, autonomy may look different in practice than in a hypothetical scenario. She gives the example of a plane crash and the sorts of investigations after a plane crash.

Investigators try to conclude who is responsible for the action – but often, an explanation might reference a failure to properly inspect a part, an operator or group of operators who created the faulty part to begin with, the person who designed the part of the system into which the part might fit, as well as an action by a pilot or air traffic controller that somehow made the part's failing more critical than it might otherwise have been. Assigning responsibility to hold people legally and ethically accountable in such a situation becomes complicated. Noorman also notes that the usual way of attributing responsibility within military communities has been to follow a chain of command, asking who gave the order for a particular activity to occur. However, those at the highest echelons of command may not have the same knowledge as a practitioner; While they are commanding those who serve, leaders are not always fully informed as to the technical limitations of the weaponry which is being used, or the nuances involved in their use (Noorman, 2014, p. 812)

Can Machines "Think Ethically"?

A final issue in the debate about machines as actors in cyberwarfare relates to self-awareness. In their work on moral responsibility, the ethicists Fischer and Ravizza (1998) suggest that an individual needs to reflect on the morality of their actions and see themselves as capable of acting in a situation requiring moral discernment.

A contrasting view suggests that a machine can be described as engaged in moral or ethical work if it merely applies a formula or chooses from a series of laws to determine whether or not an action breaks the law or violates an ethical principle. Some analysts suggest that a machine can be programmed to "think morally" if, for example, it can calculate the utility of various actions using an algorithm fed to it by programmers and then rank actions from the most to least harmful, and then take actions to either avoid or pursue these ends. In "The Ethics of Driverless Cars," Neil McBride (2015) looks at a future situation of full autonomy – where a car could decide what route to take, fill itself with gas, bid for its insurance, and learn from its environment without any human inputs. He argues that this allows the machine to go beyond human fallibility and error from an ideological perspective. Eventually, he argues your car will be a much better driver than you ever were – since it will be able to do things like interface with other cars to "choreograph" a crash. Both cars would provide technical specifications, including their speed, and then a program could decide the "best" impact for the crash to prevent property loss and the loss of human life. From a practical perspective, one might argue that the car can act morally since it could choose to behave in such a way as to reduce human casualties. McBride asks, "Who wouldn't want that?"

However, he then begins to sound like a deontologist when he asks questions about this scenario's impacts on a community. Would humans feel displaced or ashamed when a car takes their job or expertise? In such a scenario, the human is merely a means to an end, the source of inputs like gas money or insurance. He argues that humans should not be viewed as "a dangerous annoyance to be removed from the system," concluding that there are limits to how a computer or algorithm might act as a moral decision-maker.

But Fischer and Ravizza (1998) would concede that while the driverless car is perhaps displaying judgment, it is not genuinely thinking morally since doing so involves seeing oneself as able to make an independent judgment and also weighing the various reactions that might arise from making a particular decision, including the reactions of other people. Thus, they would oppose outsourcing any moral decision-making – including moral decision-making in warfare and cyberwarfare – to a machine or a bot.

And some analysts, like Sparrow or Asaro, would argue that while it might be acceptable to allow a computer to decide whether or not to run over a pedestrian, the ultimate decision over whether or not to kill in warfare still should belong morally only to the human warfighter. They believe that warfare fundamentally differs from other activities and that human interference and cognition are essential (Sparrow, 2017).

What Is War?

According to some analysts, if an AI does not understand its role in taking human life, it also does not understand the meaning of war itself. Solovyeva and Hynek (2018) raise concerns that wars themselves have a meaning that is inadequately addressed using merely quantitative data about each side's relative strengths and weaknesses. The concern is that an AI might understand a mission such as "craft a military strategy for our nation which would allow us to win a protracted war against an enemy with a larger population but fewer technological advantages." However, an AI will never fully comprehend why humans go to war, what war means to society, and how war has psychological, moral, and historical meanings for individuals who participate in it.

As Bodington notes, virtue ethics rests on the ability to make moral judgments. For this reason, he argues that an AI can't act "virtuously." He writes that "a machine, however sophisticated, would be unable to do the right things for the right reason and in the right manner" (quoted in Saveliev & Zhurenko, 2020, p. 12). AI is seen as incapable of making decisions about social responsibility, and ethicists differ on whether an AI can display an "ethic of care." Due to the limitations of even the most sophisticated AI, one can argue that AI should not be entrusted with making a decision such as "Should our nation go to war against this enemy at this time?"

Throughout history, the world's major religions – from Hinduism to Islam, Christianity to Judaism – have all produced sacred books which include scenes and themes of war and battles, including battles between those with different ideas and identities. These sacred texts describe why men fight and, in many cases, paint the activity as heroic. In this way, war is not merely a series of strategic activities aimed at maintaining the boundaries of one's territory that a sophisticated algorithm could resolve. Instead, war is conceived of as an activity with symbolic and moral importance.

In this view, wars are not merely a contest of military strength in which two sides are pitted against one another. Instead, wars are about justice, national identity, and standing up for what is right. For this reason, we write novels and make movies about wars. And this is why we memorize and recite speeches such as Prime Minister Winston Churchill's speech given in 1940, in which he reassures Britain's subjects that "We shall go on to the end, we shall fight in France, we shall fight on the seas and oceans, we shall fight with growing confidence and growing strength in the air, we shall defend our Island, whatever the cost may be."

In contrast to the noble words of Churchill, we can think about how an AI conceptualizes war. Ben Goertzel (2019) refers to the four tasks which an AI can do: selling (displaying ads and making recommendations), killing (defense and security), spying (intelligence), and gambling (investments). In each case, an algorithm allows an AI to sort information in particular ways, such as applying probabilistic assessment metrics to data. But as Goertzel's somewhat facetious comment implies, it does not matter what type of gambling an AI is engaged in with its probabilistic risk assessments. A law enforcement organization could also deploy a program that helps a drug dealer to become more efficient or better identify the likelihood of being apprehended by the police. To the AI, there is no sense of working for the good guys versus the bad guys, or any sense of whether the work which is being carried out is aimed at harming or healing the world. An AI would also not recognize the ability of war to build public morale, build a national sense of unity, or bring together diverse people from different backgrounds for a common cause.

For this reason, James Baker suggests that the best solution is one in which "the machine augments human capacity with the human seeking to understand and control the machine's capabilities while adding human judgment, strategy, and intuition to the machine's operation." He notes that the ethical issues facing states in the future will be when to rely on the machine alone versus relying on machine augmentation of human capabilities, as well as how to assert positive control over technology when the machine operates at "machine speed" and often not in a transparent fashion (Baker, 2020, p. 4).

Roff raises a criticism that is both ethical and methodological. She notes that groups like the US Department of Defense-sponsored Intelligence Advanced Research Projects Activity (IARPA) support projects that allow machines to work with socially oriented data rather than purely mathematical data. She describes how the EMBERS systems rely on artificial intelligence to predict social phenomena – or how people might behave. However, she suggests that just because artificial intelligence systems might perform well in predicting, for example, "when components on an air platform might fail from youth, corrosion or heat," they should not be trusted to predict "human behavior in complex dynamic and highly uncertain systems" (Roff, 2020, p. 1). She questions whether an AI-enabled system will ever succeed in predicting how humans will behave, since humans are not molecules that, for example, spread out when the air temperature around them rises. In addition, she argues that a human can't give direction to an algorithm or provide information to a machine-learning system about human behavior which is purely objective or neutral, since theories about human behavior tend to incorporate researchers' own beliefs about what humans do, what humans want, and what humans value. She writes that programming decisions "are not value-neutral even if they rest . . . on widely accepted theories in social science." Here, she cautions that such theories are developed in western societies and may rest on western assumptions (for example, identifying a relationship between a participant's educational level and the likelihood of participating in an insurgency). She also asks whether models that rely primarily on past events can be sufficiently predictive of future events.

Even if we were to succeed in solving the methodological issues she raises, many still believe that an overreliance on automated warfare raises the specter of an emotionless, bloodless war – one begun and conducted without human emotion and military honor. In his argument for the outlawing of lethal autonomous weapons, Sparrow describes a future scenario in which autonomous weapons might kill large numbers of human opponents without being fully aware of the fact that they are taking human lives as a form of "extermination." He argues that each human deserves the right to have another human recognize their death, rather than being destroyed by an unrecognizable, nameless agent of destruction (Sparrow, 2017).

Many years ago, an AI which was carrying out the task of identifying and classifying photos misclassified a weather pattern map as a panda bear, based on an algorithm designed to plot where the shades of color appear in an image and then measure distances between various color plots to make a reasonable estimation of what an image might be. To the AI, it didn't matter whether it was a weather map or a panda bear, though a human viewer would quickly distinguish between a living animal and a chart containing data points. In the same way, a human would participate in war differently, react differently to the events of the war, and learn different lessons from the war. Therefore, many argue that war is too important to be left to machines or solely to machines.

Here, a related virtue ethics argument can be raised by asking, "Do we want to be the sort of society which exterminates our enemies in an emotionless war that neither recognizes their humanity nor accords them the dignity of an honorable death, instead merely extinguishing them like something which is not fully human?" In short, could a fully automated war ever be virtuous or honorable?

A Utilitarian View

The main argument against autonomous weapons put forth by those who hold a utilitarian view concerns not the meaning of actions or how others might be affected. Instead, utilitarians are concerned with the risks of relying on these weapons and the likelihood that problems could crop up.

Here, programmers have raised concerns about robots' lack of situational awareness and rigidity, pointing to an AI's inability to adapt its programming rules to a dynamic, changing environment. (These specific weaknesses could also create related ethical dilemmas – if, for example, a weapon is not satisfactorily capable of distinguishing between civilian and military targets, leading it to kill civilian schoolchildren, for example, rather than suspected terrorists.)

Other concerns include the possibility of bugs and vulnerabilities and that a product's "beta version" is seldom wholly accurate compared to later versions. New technology frequently contains unknowns. In thinking about autonomous weapons, programmers have also raised concerns about the unpredictability of device–device interactions. Here, Salevyev and Zhurenkom ask what might happen when a mobile device controlled by software programs interacts with a competing hostile device controlled by unknown software? They tell us that "the result of the interaction is scientifically impossible to predict" (Saveliev & Zhurenko, 2020, p. 12).

Baker (2020, p. 3) raises concerns about nations' tendency to cut corners, mainly when China, Russia, and the United States are engaged in an AI arms race:

> Where there is parity, states will seek the marginal advantages of additional speed, additional data, and additional autonomy, and they will take shortcuts in time and safety to do so. This is what happens in technology races when security is at stake.

Other risks identified with the advent of this technology include the risk that AIfAi could cause global instability or enhance the power of authoritarian regimes.

Finally, some ethicists apply an older argument that has arisen whenever a new military technology is introduced. Pacifists (those who feel that nations should never engage in military conflicts) believe that new technologies inherently make war "less unthinkable." New technologies tend to lower the costs of war and make it easier and faster to mobilize for war. In this way, they argue, new technologies often lower the threshold for engaging in war, making it more likely that such conflicts will occur. When thinking about artificial intelligence and its role in warfighting, analysts thus worry that someday going to war may be as simple as giving a command or pushing a button. In such a situation, people may go to war for the wrong reasons or at the slightest provocation, rather than being forced to think about what war entails – as you might if you were physically loading ammunition or driving a vehicle to a conflict site.

Another utilitarian argument against AWS focuses on strategic considerations. Some ethicists are worried that as superintelligence outpaces our human intelligence, we may find ourselves in a situation where we lose human control over how wars start, escalate, and finish. In short, we may not understand "our strategy" if it is crafted for us by an AI and implemented by an AI.

Should an AI Decide When to Go to War, and Why?

An AI could tell us if we are likely to prevail in scenarios that might be simulated before engaging in actual combat. Therefore, it could tell us not to go to war if a scenario appears fundamentally unwinnable. But should an AI decide whether to go to war? Is this the type of authority we humans wish to cede to AI? How much and to what degree do we trust AI to make this decision?

Beyond the inability of AI to perhaps understand the myriad, often non-quantifiable reasons why a society might go to war, AI might also struggle to engage in war ethically – hewing to the norms of war and considering non-strategic issues such as international humanitarian law. As noted at the beginning of this chapter, today, there is often no clearly defined battlefield, and states struggle to maintain their sovereignty or control over their territory in a world where an adversary government can speak directly to citizens of your nation through video communications and social media. And earlier in this textbook, we introduced the notion of brain hacking. Can you imagine a future in which an adversary might hack into your government's computer systems and change data or use social media to alter public opinion – either to convince your nation to enter into a conflict or to dissuade them from doing so based on a belief that defeat was inevitable? An adversary might also change data in, for example, the Ministry of Defense, about the assets available and their state of readiness. However, there are no explicit norms spelling out whether such actions would be regarded as ethical or lawful.

Can AWS Observe and Understanding International Humanitarian Law?

Indeed, one might argue that it is an impossible task to "socialize" an AI into respecting and understanding, as well as applying, the norms of war. As noted earlier, war is not merely about testing one set of weapons systems against another. Instead, it is an activity that involves societies, warfighters, and civilians. Over hundreds of years (generally dating back to the Middle Ages), human societies have developed norms or shared understandings about ethical and unethical conduct during war. Thus, some ethicists are worried that since AWS can't fully understand these norms or their creation, they will likely undermine and destroy them if they enter into warfare as planners, leaders, or combatants.

Traditionally, international law, including the *Law on Armed Conflict* (*LOAC*) and the *Geneva Convention*, has provided the basis for discussing conflict ethics – including what acts are deemed lawful or unlawful. International law is written at the state level and guides states' actions in prosecuting or carrying out the war. It deems some actions lawful and legal while others are not, based on a set of ethical principles known as Just War. The Geneva Convention provides for the rights and safety of those who are not or no longer engaged in war – including civilians and those taken as prisoners of war. Although these legal understandings state what actions a state might rightfully and ethically take during combat, they are also translated nationally into policy guidance for uniformed, professional soldiers. Thus, like a state, individuals can be prosecuted for a war crime if they violate international laws and norms by engaging in an activity like torturing a prisoner or conducting military aggression against unarmed civilians. In cyberethics, the challenge is to translate these understandings to judge the ethics of conducting operations in cyberspace. What might it mean, for example, to say that someone has conducted an act of aggression against an unarmed civilian in cyberspace? We will explore this issue in greater depth in this chapter.

As Solovyeva and Hynek (2018) note, there is frequently tension between warfighters' strategic and tactical goals and international humanitarian law. (That is, what an army wants and needs to accomplish may be deterred by such factors as refugees and civilians in harm's way.) Navigating this tension has traditionally required exceptional human leadership skills to negotiate. Do we believe that AI and machine learning algorithms will be up to this task? Many ethicists do not.

A machine might also fail to observe (or be incapable of understanding) distinctions regarding when deception is morally acceptable as a "ruse of war" and when deception constitutes an ethical breach, for example. In her work, Roff raises the issue of betrayal as a type of harm that might occur in a war involving artificial intelligence. Perfidy is a specific type of criminal activity during the war when a warfighter pretends to be a civilian or pretends to surrender to advance upon a target. Such an act is seen as criminal as it helps to erode the boundary between civilians and those who are professional soldiers. In effect, the warfighter is cheating because they benefit from the protections accorded to civilians during wartime and use them for military gain. Roff suggests that the attribution problem, or the fact that sometimes it is difficult to identify a specific source from which an attack is emanating, provides a way for cyber participants in a conflict to engage in treason (Roff, 2016). Someone might open an e-mail believing it is from a civilian electricity provider, for example, and download malware created by a military person as part of a military attack. While international law and the Law of Armed Conflict do allow states to carry out certain types of ruses or "strategies of war" (such as a using a decoy ship to make your enemy believe that you have more resources than you do), impersonating a civilian during wartime is never allowed as part of lawful deception. However, one can conceptualize a situation in which, for example, an autonomous agent might decide to use the attribution problem to "impersonate" a civilian AI. Theoretically, an AI could use voice-generating algorithms to mimic the voice of a commander to issue false orders or pretend to be a member of the opposing army's cyber capabilities. Such behaviors can erode the norms of war that have been developed over hundreds of years.

In his work, *Our Final Invention*, Barrett (2013) makes an even more sobering and pessimistic observation. He believes that it is possible for a superintelligent AI to form an intent (versus merely carrying out a program to, for example, identify images in photographs) – but he believes that we cannot assume that the intent that an AI would have would be the same as ours. He suggests that we are mistaken in assuming that an AI and a human would necessarily understand their interests to be the same. He describes algorithms as inscrutable and raises a specter of how AI might act and even act with an evil intention rather than a compassionate manner.

In thinking about a situation in which we might "draft" an AI to fight in a war, the AI directing actions or strategy might either misunderstand (or misinterpret) our intentions when asked to carry them out, or form their different intentions from ours. In this future, AI might go to war for reasons we do not understand (because they are both mysterious and opaque and operate according to a different calculus than we do) or actively seek to harm us. Barrett would likely suggest that, therefore, we have no reason to believe that superintelligent AI would either accept our norms and agree not to modify them or that they would not actively seek to undermine our norms. (That is, a superintelligent AI might reject a norm that we have developed over hundreds of years, such as the fact that warfighters attempt to keep societal disruption to a minimum and do not actively cause or seek to increase refugee flows in a region.)

On the other hand, norms evolve and change, often resulting from new technological developments. Today, some norms of warfare are already being questioned – for example, how meaningful are distinctions between civilian and military spaces, or between the village and the battlefield? Today, someone in one nation can fire a drone at someone in another country using autonomous launching platforms. This way, the norm that "only professional soldiers on a battlefield may participate in war" is already questioned. And some ethicists argue that war is no longer adequately understood as an ethical contest in which joint injury is possible – if one combatant is shielded from injury by sitting behind a computer screen. In this way, one might argue that even if AI does modify the norms of war, this is neither unexpected nor necessarily an ethical risk.

Box 10.1 Going Deeper: The US Space Force and the Threat of a Hacker Targeting a Space Satellite

What do you see when you look up in the night sky? Increasingly, your view may include stars and planets and a growing number of space satellites. As of 2020, approximately 2700 operational satellites were orbiting the earth. Of these, approximately 1000 were commercial or military satellites for carrying data like satellite communications, approximately 500 were used to observe the earth for scientific purposes, and approximately 100 provided navigation and geolocation services. In addition, approximately 3000 defunct or decommissioned satellites are still orbiting the earth.

The nations with the most significant presence in space include the United States, Russia, China, Japan, and the United Kingdom. Increasingly, satellites are designed, launched, and managed by commercial entities such as Amazon and SpaceX. Satellites provide data for diverse activities, including urban planning, agriculture, and border security.

New Space vs. Old Space

Launching a space satellite is now cheaper and simpler than ever. The small size of some satellites means that multiple satellites can be launched into space in one space flight, and satellites can now be launched as constellations or even mega-constellations. A mega-constellation is a group of up to 10,000 satellites that can communicate with a ground station on earth and with one another.

Satellites today are increasingly automated and autonomous. Satellites can perform scheduled maintenance functions without a human in the loop. In the future, artificially intelligent systems may decide how to respond to threats such as a possible collision with another satellite without human interference. The communication software that runs a satellite can include open-source code and may often be available as an off-the-shelf software purchase (Poole et al., 2021).

While it is relatively simple to launch and create a space satellite, it is also relatively cheap to target a space satellite. For this reason, satellites, including high-tech mega-constellations, are vulnerable to asymmetric threats from even low-level hackers. Defending a satellite is significantly more expensive and challenging than carrying out offensive actions against a satellite. To cite an example, in 2007, hackers could take control of satellites belonging to the US National Aeronautics and Space Administration (NASA) for ten minutes (Kan, 2022).

Threats to Space Security

As noted, many functions in society are highly dependent on satellites. Satellites can carry signals that affect everything from airplanes' navigation, to financial data storage for significant corporations, to operation functions like electrical grids. Space satellites thus form part of a nation's critical infrastructure, which must be protected from attack in peacetime and wartime. But how secure are these systems, and how likely is an attack on a nation's commercial and state-owned satellites?

Satellites are best understood as a set of components that include a ground station, a communications package that allows the people on the ground to talk to and interact with the satellite, and the satellite itself, including its software and hardware. Satellites have software to run their internal functions and communicate and interact with other satellites. Each of these components faces unique security threats and presents risks to overall security.

Threats to the ground station may include the risk of physical attacks and unauthorized access. The communications systems which allow those on earth to interact with and speak to satellites are particularly vulnerable to cyber threats – including jamming, eavesdropping, jacking, spoofing, and phishing. The space vehicles themselves are vulnerable to breaches, the exploitations of software vulnerabilities, DDoS attacks, and the possibility that they could be used for criminal activity or even destroyed (Manulis et al., 2021).

In the United States, guidelines for the security of space satellites are provided by the US Department of Defense and the US Department of Commerce's National Institute for Standards and Technology. The 2020 Space Capstone publication *Spacepower Doctrine for Space Forces* lays out the responsibilities of both state and commercial actors in the areas of design and production of space satellites, while the NIST publication provides more specific standards for how commercial entities should think through and prepare for the risks they are liable to face in the space arena.

However, neither entity has issued legally binding regulations regarding how commercial actors, in particular, are to respond to threats. In addition, commercial entities in the United States are not, at this time, required to report cyber-attacks or breaches on their systems to a central government reporting entity (like a Cyber Incident Response Team. or CIRT) – as other providers of critical infrastructure are.

Satellite Attacks as Part of Warfare

The vulnerabilities presented by a nation's commercial and state-level dependence on satellites are ripe for exploitation, particularly in warfare. In February 2022, a group of hackers gained remote access to satellite internet infrastructure belonging to the international company Viasat. Most cyber analysts believe that the attack was the work of Russia, as it sought to lay the groundwork for the Russian invasion of Ukraine. The malware, dubbed Acid Rain, was designed to perform an in-depth wipe of both file system and storage device files on an infected modem. The malware then triggered a reboot, rendering user systems inoperable. Computers throughout Ukraine and Western Europe were affected (Kan, 2022).

While Russia is clearly to blame for carrying out the attack, some cyber analysts suggest that satellite providers bear some responsibility for failing to secure their systems against such attacks. In the future, wartime tactics might include scenarios in which nations target and destroy one another's satellites. For this reason, the United Nations is discussing whether all states should sign an agreement or treaty pledging not to destroy each other's satellites (Porras, 2019).

Bibliography

Akoto, W. (2020, February 25). To secure satellites, bolster cybersecurity standards in space. *Undark.org*. Retrieved from https://undark.org/2020/02/25/satellite-cybersecurity-standards/.

Kan, M. (2022, March 31). Viasat hack tied to data-wiping malware designed to shut down modems. *PC Magazine*. Retrieved from https://www.pcmag.com/news/viasat-hack-tied-to-data-wiping-malware-designed-to-shut-down-modems.

Manulis, M., Bridges, C., & Harrison, R. S. (2021). Cybersecurity in new space: Analysis of threats, key enabling technologies and challenges. *International Journal of Information Security*, *20*(3), 287–311.

Paganini, P. (2013, September 18). Hacking satellites . . . look up to the sky. *Infosec.org*. Retrieved from https://resources.infosecinstitute.com/topic/hacking-satellites-look-up-to-the-sky/.

Poole, C., Bellinger, R., & Reith, M. (2021, September 22). Shifting satellite control paradigms: Operational cybersecurity in the age of megaconstellations. *Air and Space Power Journal*, *35*(3), 46–56. Retrieved from https://www.airuniversity.af.edu/ASPJ/Display/Article/277761/volume-35-issue-3-fall-2021.

Porras, D. (2019). Anti-satellite warfare and the case for an alternative draft treaty for space security. *Bulletin of the Atomic Sciences*, *75*(4), 142–147.

Satter, R. (2022, March 15). Satellite's outage caused "huge loss in communications" at war's outset – Ukrainian official. *International Business Times*. Retrieved from https://www.ibtimes.com/satellites-outage-caused-huge-loss-commnications-wars-outset-ukraine-official-3438071.

World Economic Forum. (2020, October 23). *Who owns our orbit: Just how many satellites are there in space?* Geneva: Author. Retrieved from https://www.weforum.org/agenda/2020/10/visualizing-earth-satellites-space-spacex/.

A Utilitarian Argument for the Use of AWS?

Despite the concerns raised here – the inability of an AI to observe and comment on the moral import of its actions and the possibility that it would seek to go around rather than observe norms like international humanitarian law – not all utilitarians are opposed to an increased reliance on AWS. While one might argue that AWS might generate potential harm, one could conversely argue that such weapons are often more skilled than humans, even in the art of warfare. Gvosdev argues that cyber weapons could be created which were more precise – better able to discriminate against targets, and thus less likely to lead to indiscriminate killing:

> Whereas a generation ago, a nation might still have to use munitions to cripple strategic targets such as command-and-control facilities, power stations, or communications centers – running the risk of killing employees and personnel in those areas; a cyber weapon offers the prospect of a bloodless strike, of being able to take infrastructure offline without having to destroy it permanently . . . A cyber weapon might be seen as a more ethical choice than deploying an explosive.
>
> (Gvosdev, 2014, p. 1)

However, Gvosdev notes that one might need to plant cyberweapons on an enemy's computer during peacetime, leaving them to deploy in a war. However, as we know, doing this would violate international ethical and legal standards, which do not permit moving weapons into place – onto another state's territory – during a time of peace and without a declaration of war. Thus, Gvosdev argues that the use of cyberweapons could easily lead to a merging of the boundaries between peacetime and wartime and create a situation where even when nations were at peace, it was only because they were passively threatening their opponents with the threat that they could destroy them at any time. He also notes that it is unclear what we should call the people who create these weapons and place them upon other's servers during peacetime. Are they combatants, noncombatants, or something else? In many ways, cyberwarfare is creating a new type of war.

Domains of Warfare: What Is the Cognitive Domain?

At the beginning of this chapter, we suggested that under certain circumstances, cyberattacks should be understood not as acts of war but as a facet of psychological operations. Psychological Operations, or PSYOP, are planned operations to convey selected information and indicators to audiences to influence their emotions, motives, objective reasoning, and ultimately the behavior of organizations, groups, and individuals. Psychological operations refer to activities performed during the war and in a conflict. Here, the aim is often to "prepare the battlefield" by confusing actors as war participants by supplying false information. Psychological operations are aimed at eliciting an emotion such as confusion or fear. They may not involve physical force or the intended targets' injury (Unver & Ertan, 2022).

Psychological operations are thus a form of what Brey (2007, p. 27) calls information warfare. He defines information warfare as "an extension of ordinary warfare in which combatants use information and attacks on information and information systems as tools of warfare." Information warfare might include using the media to spread propaganda. It might also include disrupting, jamming, or hijacking the enemy's communication infrastructure or propaganda feeds and hacking into computer systems that control vital infrastructure.

Information warfare relies on deception, which is legal according to the Laws of Armed Conflict, which permits deceiving one's enemy through stratagems and ruses. Indeed, the history of warfare contains many examples of situations in which a nation successfully used propaganda as part of its arsenal of tools in warfare. In the US military, propaganda refers to "any form of communication in support of national objectives designed to influence the opinions, emotions, attitudes or behavior of any group to benefit the sponsor, either directly or indirectly" (Garrison, 1999, p. 4).

However, while deception in warfare is not considered illegal, ethicists often describe it as unethical. Psychological operations frequently aim to destroy an organization's credibility, a corporation, or an individual – including a political figure – or a process (like the electoral process). Thus, one can list objections to using cyber psychological operations within warfare from virtue ethics and deontological perspectives. A virtue ethics perspective might suggest that while warfare is honorable under certain circumstances (Vallor, 2013), deceit and deception are dishonorable in wartime activities. A deontologist would instead focus on how acts of deception frequently seek to humiliate or embarrass an opponent, harming him psychologically, if not physically. Such actions require treating an adversary (and the adversary's information) as a means to an end rather than an end in itself.

Disinformation and the Cognitive Domain

In this chapter, we have considered the brain a battlespace and a contested domain due to the advent of cognitive warfare. In this new type of conflict, many of our traditional boundaries, which we have placed around warfare, are no longer relevant. Traditionally, wars have been fought within a geographically bounded area (the battlefield) by professional warfighters (not civilians), for clearly specified military objectives, within a specified period. In contrast, cognitive warfare appears to be without borders or boundaries, does not respect the distinction between civilians and warfighters, does not appear to serve a clear military objective, and has been described as ongoing and never-ending. Such a new type of warfare raises numerous ethical challenges – both for those who would participate as warfighters and those attempting to secure information as cybersecurity professionals.

In 2020, The North Atlantic Treaty Organization (NATO) described cognitive warfare as "the weaponization of brain sciences" through exploiting the human brain's vulnerability to implement more sophisticated social engineering. A 2020 NATO report stressed that "the brain will be the battlefield of the twenty-first century. Humans are the contested domain" (Norton, 2021).

In other words, established means of psychological warfare have become more effective than ever in the new information environment. As a result, both civilians and military members must defend themselves by not falling prey to online disinformation and misinformation. Bodine-Baron et al. (2018) describe Russia's disinformation strategy as a chain with four parts. Ideas are created at the top echelons of government by political and military leaders (link 1). These orders are sent down to various organs and proxies like the Internet Research Agency, who refine the narratives to be "sold" to information consumers (link 2). Then narratives are placed on "amplification channels" such as social media (link 3). Finally, they are both picked up and disseminated further by information consumers who are most likely unaware of where the information began or its purpose (link 4).

Misinformation and disinformation narratives have two goals: Broad or diffuse narratives are propagated on social media in alarming volumes and at alarming rates. Their purpose is to disrupt societies by creating a "firehose of falsehood" that pollutes the information environment, making it harder for individuals to tell the truth from falsehood and causing them to lose faith and trust in the media environment. At the same time, targeted narratives are created, designed to appeal to a preselected group of consumers. (For example, individuals who already support anti-vaccination groups might be targeted to receive specific information designed to cause them to refuse vaccination against COVID-19. Those who have already "liked" one or more conspiracy theory-type websites online might be targeted with messages incorporating conspiracy theories about COVID-19.) Here, the aim may be to cause people to become more polarized in society by exploiting preexisting social cleavages and drawing people into committing specific actions – like targeting their opponents for violence.

Currently, military strategies do not address the cognitive domain as a theater of war, although they discuss the traditional domains of space, cyber, air, sea, and land. But, as Schmidt (2020) notes, cognition has always been important, going back to the Chinese military strategist Sun Tzu, who wrote about winning wars through intelligence, information, and deception:

> Chinese military doctrine recognizes the importance of the cognitive domain, particularly in pre-conflict or crisis stages in the continuum of conflict period. In the initial stages of the conflict, deception and unconventional "attacks" in the cognitive domain will shape how adversarial populations think. They will target a nation's human capital. They will target societal weaknesses and social networks in cyber and information systems.

A Final Word

Practitioners in the field of cybersecurity will play a vital role in the conduct and execution of cyberwarfare activities – on both an offensive and a defensive level. In Chapter 1, we introduced the idea of an ethical "grey zone," particularly when there is not yet an established ethical consensus. As this chapter has shown, fast-moving technological advances have opened up such an ethical grey zone in the area of cyberconflict. Practitioners and analysts disagree about which actions should be morally allowable and which should be prohibited, as well as the limits to which both human and AI actors should adhere. Rules which were previously a matter of conventional morality (such as the understanding that one should not target civilians in wartime) have been eroded with the discovery of the cognitive domain and the advancement of disinformation techniques. Nonetheless, as shown in this chapter, the three frameworks – virtue ethics, deontological ethics, and utilitarian ethics – can still play a valuable role in helping practitioners to think through the ethical issues they encounter, even in the absence of a moral consensus at this time.

CHAPTER SUMMARY

- Today, the borders between peacetime and wartime, between military members and civilians, and between geographic borders and sovereignty are being eroded due to technological advances.
- Today, civilians may be targeted during peacetime and wartime in the cognitive domain through disinformation techniques that seek to undermine their trust in government and sow societal confusion.

- Currently, artificially intelligent agents are deployed in warfighting activities at various levels of autonomy. Policymakers and ethicists differ about whether AI-enabled machines should ever act fully autonomously in wartime.

DISCUSSION QUESTIONS

1 **Elections and Cyberwarfare**
Go to https://help.twitter.com/en/rules-and-policies/election-integrity-policy (accessed June 21, 2022) and read Twitter's 2021 Election Integrity Policy. Focus on the activities that Twitter describes as not violating the rules about interfering in an election. Think about actions on that list that you might have engaged in – such as sharing a meme about a candidate in an election.

   - What are some of the challenges that someone engaging in platform moderation – either manually through screening posts or writing code to screen posts might encounter?
   - Go online and find some examples of posts that you view as potentially problematic using the rules posted by Twitter.
   - Have you ever reported a post that you found problematic? Share the example with the class.

2 **Targeting Another Country's Data**
In *The Centaur's Dilemma*, Baker (2020) writes that "Because most AI research and development is taking place in academic and corporate laboratories, these facilities are intelligence targets in ways they have not been before." As a result, he writes, "We can expect federally funded research and development centers (FFRDCs), university research centers in corporations like Google and Facebook to become perpetual adversarial targets."
Baker implies that most private enterprises will not have either sufficient motivation or skills to counter the efforts that adversary nations might be willing to make. He also notes that hostile nations and hostile actors like terrorists might seek to collect data for machine learning – including genomic data and health records.
Given the concerns that Baker articulates:

   - Should such data be safeguarded and held by the government, not private industry?
   - Do you feel that the government has an *obligation* to step in and safeguard this data?

   Here, you again see a clash between two competing ethical values: the desire to preserve democracy, prevent autocracy, and safeguard individuals' data.

   - Which approach would you choose, and how might you ethically justify your decision?

3 **Ethics of Cognitive Warfare**
In this chapter, we have considered the brain a battlespace and a contested domain due to the advent of cognitive warfare. In this new type of conflict, many of our traditional boundaries, which we have placed around warfare, are no longer relevant. Traditionally, wars have been fought within a geographically bounded area (the

battlefield) by professional warfighters (not civilians), for clearly specified military objectives, within a specified period.

In contrast, cognitive warfare appears to be without borders or boundaries, does not respect the distinction between civilians and warfighters, does not appear to serve a clear military objective, and has been described as ongoing and never-ending.

- What ethical challenges does this new type of warfare raise for those who would participate in it as warfighters and those who secure information as cybersecurity professionals?
- What specific values should a cybersecurity professional prioritize in participating in or responding to cognitive warfare?

4 **The Problem of Espionage and Intelligence**

Cunliffe (2021) describes the threat of "street surveillance" for human intelligence sources due to advanced CCTV networks. He describes how an individual arriving at customs at the airport in Moscow is likely to go through several biometric checkpoints that might collect information like his fingerprints, iris scan, and facial scans. Biometrics (EU) is used for immigration and law enforcement.

He suggests that counterintelligence may include injecting malware into foreign biometric systems and corrupting or destroying data to safeguard the identities and persons of those engaged in espionage against the host country.

- Do you agree with Cunliffe about the necessity of destroying the biometric data banks of adversarial countries?
- What would be the ethical pros and cons of doing so?
- Would you consent to be part of such an operation, making you an "ethical hacker"? Why or why not?
- See if you can apply the three ethical lenses (utilitarianism, virtue ethics, and deontological ethics) in formulating a response to this question.

5 **Understanding Disinformation**

Watch Alexa Pavliuc, January 6, 2020, "Watch Six Decade Long Disinformation Operations Unfold in Six Minutes," Medium.com. Available at https://medium.com/swlh/watch-six-decade-long-disinformation-operations-unfold-in-six-minutes-5f69a7e75fb3 (accessed June 21, 2022).

This article and the accompanying video presentation show how tools can be used to map hashtags and accounts on social media and the connections between hashtags and accounts. What does this presentation add to your understanding of the evolution of cyberwarfare tactics like psychological operations and disinformation?

RECOMMENDED RESOURCES

North Atlantic Treaty Organization Innovation Hub Allied Command for Transformation. "cognitive warfare Project – Reference Documents." (2020). Retrieved May 6, 2021, from www.innovationhun-act.org/cw-documents-0.

United States Office of Science and Technology Policy. (2020). *National strategy for critical and emerging technologies*. Retrieved from https://trumpwhitehouse.archives.gov/ostp/documents-and-reports.

References

Baker, J. E. (2020). *The centaur's dilemma: National security law for the coming AI revolution.* Washington, DC: Brookings Institution.

Barrett, J. (2013). *Our final invention: Artificial intelligence and the end of the human era.* New York: Thomas Dunne Publishers.

Bodine-Baron, E., Helmers, T., Radin, A., & Tryger, E. (2018). *Countering Russian social media influence.* Santa Monica, CA: Rand Corporation.

Brey, P. (2007). Ethical aspects of information security and privacy. In N. A. Petkovic (Ed.), *Security, privacy and trust in modern data management* (pp. 21–36). Heidelberg; Berlin: Springer.

Chester, D. (2020, January 14). How and why power grid cyberattacks are becoming terrorists' go-to. *Energy Central.* Retrieved from https://energycentral.com/c/iu/how-and-why-power-grid-cyberattacks-are-becoming-terrorists-go.

Cunliffe, K. (2021). Hard target espionage in the information era: New challenges for the second oldest profession. *Intelligence and National Security, 36*(7), 1018–1034.

Dowdy, J. (2018, June 11). AI and autonomy: Trying to piece it all together. *Linkedin.* Retrieved June 1, 2022, from www.linkedin.com/pulse/ai-autonomy-trying-piece-all-together-john-dowdy/.

Fischer J. M., & Ravizza, M. (1998) *Responsibility and control: A theory of moral responsibility (Cambridge Studies in Philosophy and Law).* Cambridge: Cambridge University Press.

Floridi, L. (2013). *The ethics of information.* Oxford: Oxford University Press.

Garrison, W. (1999). *Informaiton operations and counter-propaganda: Making a weapon of public affairs.* Carlisle Barracks, PA: US Army War College.

Goertzel. B. (2019, July 31). Will artificial intelligence kill us? *YouTube.* Retrieved from www.youtube.com/watch?v=TDCIKEORtko.

Gvosdev, N. (2014, January 30). The ethics of cyberweapons. *Ethics and International Affairs.* Retrieved from http://www.ethicsandinternationalaffairs.org/2014/the-ethics-of-cyberweapons/.

Hildreth, S. A. (2001). *"Cyberwarfare" Congressional Research Service, FL30735, 16.* Retrieved April 14, 2017, from https://fas.org/sgp/crs/intel/FL30735.pdf.

Kirkpatrick, D. (2017, November 15). Signs of Russian meddling in Brexit referendum. *New York Times.* Retrieved June 1, 2022, from www.nytimes.com/2017/11/15/world/europe/russia-brexit-twitter-facebook.html.

Manjikian, M. (2014). Becoming unmanned: The gendering of lethal autonomous warfare. *International Feminist Journal of Politics, 16*(1), 48–65.

Matthias, A. (2004). The responsibility gap: Ascribing responsibility for the actions of learning automata. *Ethics and Information Technology, 6*(3), 175–183.

McBride, N. (2015). The ethics of driverless cars. *ACM SIGCAS Computers and Society, 45*(3), 179–184.

Miller, K., Wolf, M., & Grodzinsky, F. (2015). Behind the mask: Machine morality. *Journal of Experimental and Theoretical Artificial Intelligence, 27*(1), 99–107.

Noorman, M. (2014). Responsibility: Practices and unmanned military technologies. *Science and Engineering Ethics, 20*(3), 809–826.

Norton, B. (2021, October 13). Beyond NATO's "cognitive warfare": "Battle for your brain" waged by western militarities. *MR Online.* Retrieved June 13, 2022, from https://mronline.org/2021/10/13/behind-natos-cognitive-warfare-battle-for-your-brain-waged-by-western-militaries/.

Pascucci, P., & Sanger, K. (2021, July 2). Revisiting a framework on military takedowns against cybercriminals. *Lawfare Blog.* Retrieved June 13, 2022, from www.lawfareblog.com/revisiting-framework-military-takedowns-against-cybercriminals.

Puiu, T. (2022, March 15). How open-source intelligence (OSINT) is exposing the Ukraine war in real-time. *ZME Science.* Retrieved June 6, 2022, from www.zmescience.com/science/news-science/open-source-intelligence-ukraine/#:~:text=The%20ongoing%20war%20in%20Ukraine%20is%20rather%20unique,on%20this%20humanitarian%20crisis.%20Here%E2%80%99s%20why%20that%20matters.

Roff, H. (2016). Cyber perfidy, ruse and deception. In F. Allhoff, & A. A. Henschke (Eds.), *Binary bullets: The ethics of cyberwarfare.* Oxford: Oxford Scholarship Online.

Roff, H. (2020). *Uncomfortable ground truths: Predictive analytics and national security*. Washington, DC: Brookings Institute.

Saveliev, A., & Zhurenko, D. (2020). Artificial intelligence and social responsibility: The case of the artificial intelligence strategies in the US, Russia and China. *Kybernetes*, *50*(3).

Schmidt, T. A. (2020, September 9). Cognitive superiority and winning in multi-domain operaitons. *toddandrewschmdt.com*. Retrieved from https://toodandrewschmidt.com/wp-content/uploads2020/09/achieving-cognitive-superiority.pdf.

Solovyeva, A., & Hynek, N. (2018). Going beyond the killer robots debate: Six dilemmas autonomous weapons systems raise. *International and Security Studies*, *12*(3), 166–208.

Sparrow, R. (2017, March 7). Killer robots: Professor Robert Sparrow. Retrieved from https://www.monash.edu/arts/bioethics/news-and-events/articles/2017/killer-robots-professor-robert-sparrow.

Srinivasan, R. (2021, January 7). How social media fueld the insurretion at the US capitol. *UCLA School of the Arts and Architecture Blog*. Retrieved June 11, 2022, from www.arts.ucla.edu/single/how-social-media-fueled-the-insurrection-at-the-us-capitol/#!.

Unver, H. A., & Ertan, A. (2022). *The strategic logic of digital disinformation: Offense, defense and deterrence in information warfare*. Unpublished Manuscript.

Vallor, S. (2013). The future of military virtue. In K. S. Podins (Ed.), *5th international conference on cyber conflict*. Tallinn, Estonia: NATO CCD COE Publications.

Ventre, D. (2020). *Artificial intelligence, cybersecurity and cyberdefense*. Hoboken, NJ: Wiley-Iste.

Walzer, M. (2016, June 3). Terrorism and just war. *Youtube*. https://www.youtube.com/watch?v=yZEprmC65Pk.

Glossary

Accessibility by design: The practice of referencing accessibility standards as part of the design process, the first step in developing new devices and programs.

Application Program Interface (API): A set of functions and procedures allowing the creation of applications that access the features or data of an operating system, application, or another service. (For example, an app that requires access to a user's location data from another source depends on an API.)

Autonomy: The ability to act like an individual with free will and without being coerced or controlled by another.

Authentication: The use of encryption protocols to ensure that all system users or participants in a transaction are who they say they are. Usually carried out through the issuance of certificates.

Blockchain: A system where records of transactions are maintained across several linked computers (peer-to-peer network) and simultaneously updated. Helpful in tracing transactions and determining provenance.

Categorical imperative: An ethical standard proposed by Immanuel Kant. Defined as an objective, rationally necessary, and unconditional principle that we must always follow despite any natural desires or inclination we may have to the contrary (*Stanford Encyclopedia of Philosophy*).

Centralized/decentralized apps: In centralized apps, information is stored in a central repository, and apps contact the repository to check and update information. In decentralized apps, information is stored only on the user's device.

Conventional morality: Laws or rules are considered standard in a group or society because they conform to long-held beliefs and values.

Copyright: A form of legal protection on artistic products, branding, or labels.

Critical infrastructure: Sectors whose assets, systems, and networks, whether physical or virtual, are considered so vital to the nation that their incapacitation or destruction would have a debilitating effect on security, the economy, public health, or safety (United States Department of Homeland Security).

Data governance: Setting policies or standards within an organization that specifies the rules governing how data is stored, processed, and disposed of. Data governance policies also specify levels of access or privilege for different classes of users or roles.

Data localization: Policies adopted requiring that user data uploaded or provided to the internet will be stored at a facility in the same nation that the user was located in while performing computer activities; this means that the laws regarding the safeguarding of citizen information will be in conformity and compliance with the norms where the activity took place.

Data privacy: The protections afforded to an individual's data and the right to access and transfer that data.

Data trustee: A firm or agency that is legally and ethically responsible for storing and safeguarding personal and corporate user data and making decisions regarding releasing and sharing that data with others.

Deontology: Developed by Immanuel Kant, an ethical system that is sometimes called rule-based ethics. Deontological ethics aims to identify a universal set of morally right or wrong actions regardless of circumstances or intention.

Difference Principle: Developed by John Rawls, the Difference Principle holds that inequalities in the distribution of wealth, responsibility, and power are permissible as long as they benefit the lowest strata of society (Encyclopedia.com)

Differential privacy: A principle for creating algorithms that protect the privacy of individual users while still providing valuable information about a set of users through injecting "noise" into the results such that it is difficult to link a particular response to a particular individual

Digital epidemiology: The study of public health based on the analysis of digital data about community or group health practices and patterns.

Disparate Impact: An unequal or biased effect upon a protected class (such as gender, race, gender identity) as a result of a policy or practice that appears to be nondiscriminatory.

Doxing: Revealing someone's personal information publicly, without their knowledge or consent.

Ephemeral identifier: A process for identifying a user whereby the identification code changes frequently and can only be accessed by an authorized client.

Fair use: The principle that specific ideas or products can be used without payment or licensing fees provided certain limits are adhered to.

Grey area: An area of uncertainty where the norms and laws are not yet explicit or resolved, often because the problem being addressed is new or novel.

Impact assessment: A process for analyzing long-term or significant changes that may occur as the result of a particular intervention (for example, a privacy impact assessment examines how a particular policy might affect user privacy).

Intellectual property: The notion that the originator of an idea owns that idea. They are entitled to determine how it is used and to be compensated for its use.

Informed consent: Legal consent to act with knowledge of possible consequences.

Learning management system (LMS): A system used, often in universities or businesses, to provide long-term or short-term online learning. The system collects data about user behavior while engaged with the system.

Location-based services (LBS): The information or options provided to a user varies by the user's location (such as a restaurant recommendation based on the neighborhood you are in).

Location perturbation mechanism: Techniques designed to safeguard user privacy when using location-based services by reducing the precision of the information provided about location or adding "noise" to the data provided.

Objectivist: One who takes an ethical stance that assumes that moral good is real and exists independently of culture or opinion. This stance posits that the most moral decision or outcome can be identified or discovered.

Philosophy: An intellectual discipline that asks broad questions about the nature of knowledge, reality, values, mind, reason, and logic.

Plagiarism: Claiming another person's ideas or intellectual property as your own.

Privacy by design: A design process for software and computer systems, including networked systems in which privacy protocols and processes are a central feature of the design and where privacy is considered in all stages of planning for the system or product

Privacy-enhancing technologies (PET): Methods that allow online users to protect the privacy of their personal information. It may include encryption or anonymizing protocols.

Public good: A good created from which no user can be excluded (such as the provision of clean air or water) and for which one person's use does not another's enjoyment or use of the same good.

Ransomware: Software designed to steal or encrypt data belonging to users, under threat of its destruction or dissemination if a ransom isn't paid.

Security by design: An approach to software and system configuration that emphasizes building security protocols and processes through all stages of the design process. It may include creating specific data architectures and requiring authentication for users.

Surveillance as a service (SaaS): Refers to subscription-based services (such as video-enabled doorbells and home security cameras) in which software, platforms, and actions are controlled by a commercial entity for which the user pays a fee.

Ubiquitous computing: The notion that individual and corporate users are surrounded by technologies that collect user data as these technologies are incorporated into more everyday devices. Users may be unaware of these activities, which may occur without their knowledge or consent to data sharing.

Utilitarianism: A system of ethics where the utility of actions can be measured to determine their moral quality.

User-centered design: A process where designers focus on users and their needs, often involving users in the design process – collecting information on how they perceive and use a site or application, using principles of user experience (UX) research.

Web scraping: The process of collecting data from the web for commercial or private use, often through the use of automatic bots which visit web pages and collect and aggregate information.

Transparency: A business ethic that suggests companies and other entities (like government entities) must accurately represent information, provide complete information (within limits of not compromising corporate or government security), and communicate openly and clearly about their activities and goals.

Veil of ignorance: Developed by John Rawls, the veil of ignorance is an approach to ethical decision-making in which the decision-maker is charged to act as though he does not know their position or role in a scenario. Thus, they are charged with creating an outcome that benefits all, particularly the weakest and most vulnerable actors in a transaction.

Virtue ethics: A philosophical approach that states that the most moral outcome is in keeping with an individual's values. In this approach, the goal of ethics is to develop an individual's moral character.

Index

[Note: numbers in bold indicate a table. Numbers in italics indicate a figure]